Internet Access in Vehicular Networks

Wenchao Xu • Haibo Zhou
Xuemin (Sherman) Shen

Internet Access in Vehicular Networks

Wenchao Xu
Department of Computing
Hong Kong Polytechnic University
Hung Hom, Hong Kong

Haibo Zhou
School of Electronic Science
and Engineering
Nanjing University
Nanjing, China

Xuemin (Sherman) Shen
Electrical and Computer Engineering
Department
University of Waterloo
Waterloo, ON, Canada

ISBN 978-3-030-88993-7 ISBN 978-3-030-88991-3 (eBook)
https://doi.org/10.1007/978-3-030-88991-3

This Springer imprint is published by the registered company Springer Nature Switzerland AG
The registered company address is: Gewerbestrasse 11, 6330 Cham, Switzerland

Preface

Modern vehicles generate a large amount of data for many emerging automotive applications, such as road safety, traffic management, autonomous driving, and intelligent transportation system. The Internet of vehicles are expected to enable effective acquiring, storage, transmission, and computing for such big data among vehicle users and thus can facilitate better perception of both internal and external vehicular environments as well as the status of drivers, passengers, and pedestrians. To efficiently connect mobile vehicles to the Internet and conduct massive information exchange among vehicle users and the transportation system, it is essential to evaluate the Internet access performance via both the vehicle-to-roadside (V2R) and vehicle-to-vehicle (V2V) paradigms. Besides, in order to support reliable and efficient Internet access for mobile vehicle users, it is very important to explore various spectrum resources rather than solely relying on cellular networks, such as unlicensed WiFi band, TV White Space, opportunistic V2V data path. To set up effective Internet connection, a practical signaling process between a roadside access station and vehicle users is necessary, e.g., various management frames need to be transferred to set up effective Internet connections. As vehicles move around, it is possible for them to set up opportunistic connections to share the data contents, bandwidth, and computing capacities, which can enable a variety of novel computing and communication paradigms that are beneficial to future automotive applications. In this monograph, we investigate the Internet access procedure and the corresponding analytical evaluation methods, as well as novel machine learning paradigms for reliable and robust Internet connectivity on the road. We first introduce the Internet access of vehicles and propose an analytical framework for modeling of Internet access performance via the roadside hotspots, considering the necessary Internet access procedure that comprised of association, authentication, and network configuration steps, where the access delay and throughput capacity are evaluated in drive-thru Internet scenario. We then explore the interworking of different V2X communication paradigms and study the opportunistic assistance from neighboring vehicles, which apply the V2V communication to conduct Internet data offloading upon the interworking of V2V and V2R communication, where the trade-off between the delay and throughput of the V2V assistance is analyzed. In addition, we take a close look at the wireless link management between the vehicle and Internet access stations. We investigate the V2X channel that is highly varying and thus difficult to accommodate proper modulation and coding scheme to satisfy various user quality-of-service (QoS). To deal with such issue, we apply big data analytics and show that the proposed data-driven and learning-based methods can greatly reduce the packet drop rate and thus improve the Internet access performance in terms of both access delay and transmission throughput. Several case studies are presented to examine the utility of the big vehicular data to enable the intelligent Internet access. Furthermore, to train the machine learning models among vehicle users in a distributed manner, we design efficient IoV protocols to boost the training process, including the rateless coding-based broadcasting scheme for intelligent model delivery that can enhance the process of collaborative learning among vehicles, whereby asynchronous federated learning can be conducted for mobile vehicles with high mobility and opportunistic inter-contacts. We hope that this monograph will provide inspiration and guidance on further research and development of the future Internet of vehicles.

We would like to thank Prof. Weihua Zhuang from the University of Waterloo (UW), Prof. Song Guo from The Hong Kong Polytechnic University, Prof. Nan Cheng from the Xidian University, and many UW BBCR members for their contributions in the contents of this monograph and great support to related research projects. Special thanks are also due to the staff at Springer Science+Business Media: Susan Lagerstrom-Fife, for her great help in the publication of the monograph.

Hung Hom, Hong Kong Wenchao Xu

Nanjing, China Haibo Zhou

Waterloo, ON, Canada Xuemin (Sherman) Shen

Contents

Acronyms

AAA	Authenticate, authorization, and accounting
AI	Artificial intelligence
AP	Access point
AS	Aggregation server
BAP	Broadcasting enabled asynchronous parallelization
BS	Base station
COTS	Commercial off-the-shelf
CSI	Channel state information
DCF	Distributed coordination function
DL	Deep learning
DRL	Deep reinforcement learning
DSRC	Dedicated short-range communication
FEL	Federated edge learning
FL	Federated learning
IoV	Internet of vehicle
ITS	Intelligent transportation system
MAC	Medium access control
MCS	Modulation and coding scheme
ML	Machine learning
NLOS	Non-line-of-sight
OBU	Onboard unit
RA	Rate adaptation
RSSI	Received signal strength indicator
RSU	Roadside unit
SNR	Signal-to-noise ratio
TSC	Time series classification
TVWS	TV white space

UAV	Unmanned aerial vehicle
V2I	Vehicle-to-infrastructure
V2R	Vehicle-to-roadside
V2V	Vehicle-to-vehicle
V2X	Vehicular-to-everything
VANET	Vehicular network

Chapter 1
Introduction of Internet Access of Vehicular Networks

Connected vehicles are changing the modern transportation. Based on the wireless communication between vehicles with sophisticated radio interfaces, vehicles in the mobility world can exchange information with neighbors as well as remote transportation center, which can enable the vehicle to understand both the in-vehicle status and the road situation. Based on such capability, a lot of smart road applications can be realized, including the safety-related application, intelligent transportation system (ITS) and in-vehicle infotainment, etc. The Internet access for vehicles can further extend the spatial scope and temporal range of the vehicular communication, which can help all road users to conduct both the long-term evaluation and short-time response to all situations. In this chapter, we first introduce the overview of Internet of vehicles (IoV), then we present the Internet access procedure for a vehicle to connect to a wireless access station that deployed along the roadside. We then explain the aim of the book, covering the topic of Internet access performance evaluation, data traffic offloading, Internet link management and intelligent machine learning (ML) paradigm over IoV.

1.1 Internet of Vehicles Overview

To provide the ubiquitous network connection for automobiles, traditional vehicular network, which are supported with the Vehicle-to-Everything (V2X) communication technology, are evolving to the Internet of Vehicles (IoV), which greatly expand the time scope and the spatial range of traditional vehicular networks. Literature status quo show that great progress have been made toward the robust and high performance networking between vehicles and vehicles, vehicles and roadside unit. However, the emerging automotive applications, such as autonomous vehicles, intelligent transportation system, etc., have raised new requirements and challenges for future connection vehicles [1]. To satisfy the need of future

W. Xu et al., *Internet Access in Vehicular Networks*,
https://doi.org/10.1007/978-3-030-88991-3_1

automotive applications, IoV need to focus on more aspects of the road services that beyond the traditional VANET's capabilities, which mostly deals with the road safety issues. The original V2X technology is called the Dedicated Short-Range Communications (DSRC) that developed over the IEEE 802.11p protocol revised from the original IEEE 802.11a standard. The idea is to migrate the success of the WiFi network, i.e., take advantage of its simplicity, high throughput of the IEEE 802.11 based networking to enable basic vehicular wireless connections to support the safety message exchanging. Since the publication of the DSRC in 1999, various V2X technologies have been proposed, which can be divided into two categories, i.e., the IEEE 802.11 V2X and the cellular based V2X (C-V2X).

As shown in Fig. 1.1, IoV is urged to provide widely and real-time network access for emerging automotive applications, including cooperative automatic driving, intelligent traffic control, and collaborative environmental perception, etc., which require extensive data transmission and information exchange that beyond the capability of conventional connected vehicles. To better assist future road applications, IoV has to deal with the challenges such as dynamic topological variations, extended network scale, interrupted radio connection, spectrum scarcity and computation resource limitation, etc. We focus on the Internet connection

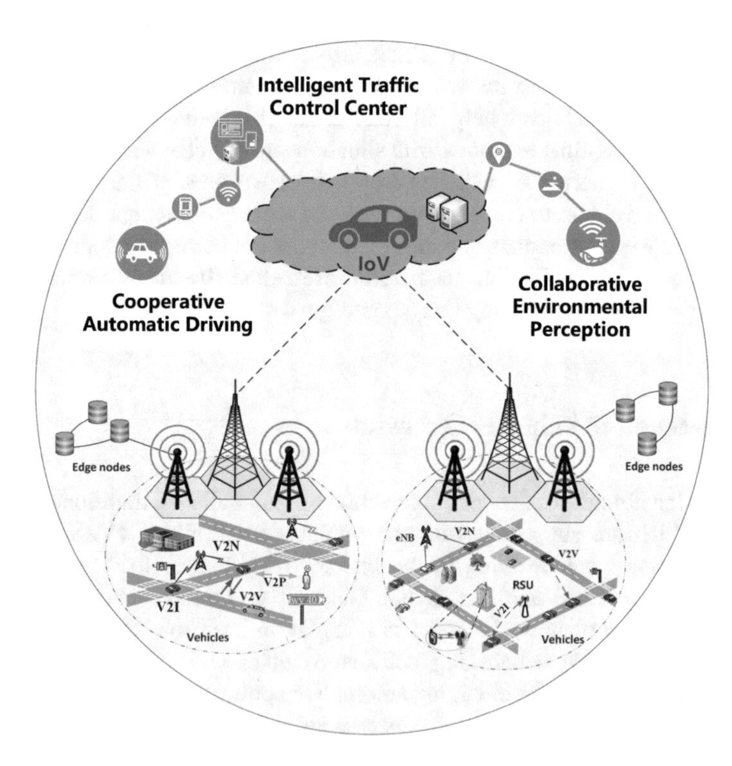

Fig. 1.1 The illustration of Internet of Vehicles

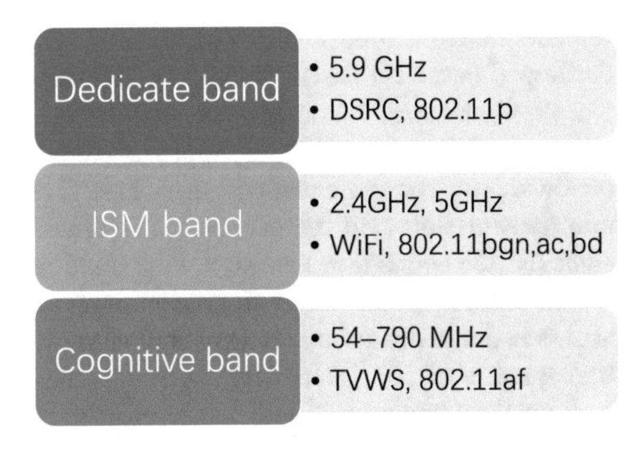

Fig. 1.2 802.11 V2X spectrum access types

Table 1.1 802.11 V2X comparison

802.11 V2X type	Access spectrum	Coverage	Penetration	Link capacity (approx.)	Cost
DSRC	Dedicated	<1 km	Poor	Less than 10 Mbps	High
WiFi	Unlicensed	100–300 m	Poor	Up to 1 Gbps (802.11ac)	Low
TVWS	Cognitive	Around 10 km	Good	10 Mbps	High

between vehicle users all road users, to verify the capability to support the transmission, computing, storage of the big IoV data.

There are two camps of Internet access technology, i.e., IEEE 802.11 IoV and cellular IoV. Compared with cellular IoV, the IEEE 802.11 IoV has attracted wide attentions due to its great advantages, i.e., to bring the success of WiFi networks to vehicular condition. There are three types of IEEE 802.11 IoV branches based on the way of accessing to the spectrum resource, as shown in Fig. 1.2, i.e., DSRC, WiFi and TVWS, whose detailed comparisons are given in Table 1.1.

1.1.1 DSRC

DSRC is the initial standard for vehicular networks that stands for the Dedicated Short Range Communications published in 1999. The standard is an amended version from the original WiFi protocol, i.e., the IEEE 802.11a standard. By revising the PHY layer parameters and fix the backoff window to prevent the excessive channel access delay. The revision is called IEEE 802.11p, which works at the 5.9 GHz band and only uses half of the IEEE 802.11a's bandwidth to support same-level modulation and coding scheme [2]. The DSRC protocol works at the ad-hoc mode, i.e., vehicles are the ad-hoc node that can communicate with each other directly via the V2V channel, and vehicles can also directly communicate with the roadside units via the V2R channel. DSRC is mandatory in north America for new manufactured vehicles after 2016. There are a lot of safety applications developed based on DSRC, including collision alert, emergency event informing, etc. [3, 4]. Both of the single-hop and multi-hop DSRC connections are utilized to enable ITS application to improve road efficiency [5]. Due to the limited bandwidth and communication range, DSRC cannot always support the high data rate transmission, and the connection is often interrupted due to the non-line-of-sight (NLOS) condition and highly dynamic network topology [6]. The access to DSRC channel would suffer significant delay due to the inefficiency of MAC layer access in dense conditions. Recent studies show that DSRC can support end-to-end transmission of 50–100 ms latency, which can support current safety applications well such as brake alert, and traffic efficiency enhancement. However, for some advanced vehicular applications such as remote driving, DSRC may be not feasible as they require more stringent transmission latency [7] (Table 1.1).

1.1.2 ISM Band WiFi with Opportunistic Access

Directly applying the ISM band WiFi to vehicular conditions is expected to continue their success on road due to its unlicensed spectrum advantage and the high performance. The utility of WiFi networks has been well proved by both measurement and analysis [8, 9]. In 2004, Ott et al. setup a 802.11b at 2.4 GHz WiFi AP on roadside to provide Internet access for the drive-by vehicles, which is referred to as the 'drive-thru Internet'. It is shown that considerable throughput can be achieved for both UDP and TCP traffic. Such paradigm has been used for many data applications, e.g., vehicle data offloading [10, 11], content caching [12, 13], data delivery [14], etc. Due to the limited coverage range, the connection to roadside WiFi networks would be interrupted. Cheng et al. analyzed the trade-off between the data task fulfillment delay and the offloading efficiency by opportunistically transmitting to a series of WiFi APs along the road when drive through the coverage areas [15]. Beside such opportunistic V2R communications, WiFi is also applied to enable the V2V communications, such as V2V content sharing [16].

The advantage is that WiFi is always evolving. First, the new generation WiFi network would always introduce new technologies to improve the link rate, e.g., channel binding, massive MIMO, higher order modulation, etc. [17]. For example, the 802.11ac WiFi is measured in both V2V and V2I communication and outperforms the legacy 802.11n protocol [18]. In addition to the data plane improvement, new control functions are introduced to the next generation WiFi. To enable smooth AP switch when the vehicle drives through a series of roadside WiFi networks, 802.11r is issued to reduce the number of management frames during the handover process [19]. The new hotspot 2.0 specification has enabled the Authenticate, Authorization, Accounting (AAA) functions to WiFi and access framework including the automatic association, secure communication, and better interworking with backhaul networks [20]. Furthermore, the cost of deploying WiFi networks is low. It is agile to build a roadside AP based on inexpensive hardware and open source software. Thus, it is even possible to deploy an array of WiFi transceivers along the roadside and take advantage of the path diversity. Experiment in [21] has shown that the link performance can be greatly improved for both UDP and TCP traffic. To apply WiFi technology to vehicular networks, there would be some issues cause of the high mobility. However, consider economical efficiency and network performance, the advantages apparently outweigh disadvantages, not to mention that future WiFi networks are keeping improving in terms of higher link rate, mobility support, and roaming ability, etc.

1.1.3 TVWS with Cognitive Spectrum Access

Both of the DSRC and WiFi networks have poor propagation range and penetration capability due to the high carrier frequency. To avoid the frequent handover and further improve the network bandwidth, the vacant TV spectrum between 470 to 790 MHz are re-utilized as they are often no longer used when Internet becomes the main source of information and entertainment. In 2014, IEEE published the 802.11af standard to support sharing the TVWS spectrum for cognitive secondary users [22]. To investigate the feasibility and check the efficiency of vehicular access to TVWS, Zhou et al. applied the 802.11af TV access system for both V2V and V2I communication guided by spectrum usage according to a geolocation database [23, 24]. The field measurements in [25] show that 802.11af transceivers can achieve beyond 9 Mbps over 6 km. There still remains critical issues for vehicular TVWS access. First is that the secondary users could not access if primary users occupy the TV spectrum, which means that the usability of TV band is not guaranteed. Since the coverage area is much bigger compared with the normal WiFi networks, number of co-associated vehicle users would likely be large, which leads to intensive media access contention and severe congestion [26]. Moreover, the cost of setting up a TVWS not only requires expensive investments, but also the spectrum permission from regulation office, which further limits its usage.

1.1.4 Cellular IoV

Cellular networks can provide ubiquitous coverage, seamless handover, high bandwidth access, reliable and secure message transmission, and thus has great potential to support enormous vehicular communication applications and services. Cellular IoV standardization can be divided into three phases. At the first stage, LTE network is used to support basic V2X applications since 2015 within 3GPP Release 14 [27]. 3GPP defines 27 typical use cases encompassing V2V, V2I, V2P, and V2N applications, and the requirements for 7 typical scenarios. It has confirmed to adopt the PC5 interface and Uu interface based LTE network to support IoV services, and enhancement from physical layer structure, resource selection and allocation, and synchronization have been developed by the 3GPP RAN work group (WG). In Rel-14, V2X is mostly to provide data transport service for basic road safety service such as cooperative awareness messages (CAM), basic safety message (BSM), or decentralized environmental notification messages (DENM), and so on. Enhancement has been involved in Rel-15 [28] to support advanced IoV scenarios [29], such as vehicle platooning, advanced driving, extended sensors, remote driving, etc. In Rel-15, the WI "V2X phase 2 based on LTE" introduces some key functionalities to support advanced IoV services in a fully backward compatible manner with Rel-14 V2X, including carrier aggregation (CA) for mode-4, higher order modulation (e.g., 64-QAM), radio resource pool sharing between mode-3 and mode-4 UEs, shorten transmission time interval (TTI), and reduction of the maximum time between packet arrival at Layer 1–10 ms (which was 20 ms in Rel-14) and resource selected for transmission. 3GPP has also started 5G NR-V2X recently [30], to evaluate the enhanced V2X services by defining simulation scenarios, performance metrics, channel modeling, spectrum, etc. Chinese telecom companies, such as Datang and Huawei, are among the main force in the 3GPP V2X standardization and development of several LTE-V2X applications [31]. From 2016, the newly established 5G Automotive Association (5GAA) and the Next Generation Mobile Networks Alliance (NGMN) [32] have developed V2X solutions to support connected cars and road safety applications.

1.1.5 Summary

Both of IEEE 802.11 IoV and cellular IoV will play an important role in future vehicular applications. We mainly consider the low-cost, high-throughput Internet access via the unlicensed 802.11 networks in this monograph. To provide Internet connection to vehicles, a vehicle should finish the Internet access procedure, to find the AP and get correct configurations to enable effective Internet connection [24].

1.2 Internet Access Procedure

To utilize current IoV radio interface to access to Internet, the access procedure include three main steps that involve the corresponding main functions, i.e., network detection, user authentication and network parameters configuration.

1.2.1 Network Detection

Traditional WiFi use beacon frames to exchange the information between the user and AP. The beacon scheme includes two mechanisms, i.e., passive beaconing and

Table 1.2 IEEE 802.11 Beacon frames

Type	Tag	Usage
Link parameters	Support rates	Physical layer rate/modulation type
	Channel status	Signal and noise level, spectrum, etc.
	802.11 radio	RTX frequency index, current data rate, timestamp, etc.
	DS-status	Mode (ad-hoc/BSS),
	HT info	20 MHz/40 MHz bandwidth
Network info	QBSS load	Channel utilization, admission capacity, etc.
	SSID	Network name
	Interworking	Network type and hotpsot 2.0 support
	RSN info	Authentication method
	EAP method	Type of EAP

proactive beaconing. In passive beaconing, the AP broadcasts the beacon frames to nearby users, including the information of link parameters, network ID, etc., which are summarized in Table 1.2 [33]. In proactive mode, user broadcast a probe request frame to search nearby APs, which will then reply the required information in the prove reply frames. Before access to the roadside WiFi network, the vehicle should detect the existence of the network via beacon frames exchange or other query protocols (e.g., the Access Network Query Protocol (ANQP) of Hotspot 2.0) [34]. This step is essential for vehicles to get the information about the wireless link parameters, e.g., 802.11 radio channel, supported rates, SSID, etc., and the backhaul information like authentication method, current load, etc. The information can also help the vehicle to select a proper nearby AP.

However, sometimes such information is not sufficient for clients to find a proper AP to associate. First, the clients have limited information about the backhaul network connected to the AP, e.g., Internet accessibility, the security level, the QoS mapping support. Secondly, according to the default AP selection policy, the WiFi clients always choose the one with the largest Received Signal Strength Indicator (RSSI) among all available nearby APs. Such AP selection policy may result in improper association, e.g., association to APs without Internet connectivity, or causing unbalance load distribution and low utilization of available APs [35]. In order to overcome these problems, hotspot 2.0 are specified that no longer rely on the SSID solely to identify a WiFi network [36]. Prior to AP association, the WiFi clients can obtain more information such as the Network Access Identifier (NAI), the operator information via multiple ways. First, new information elements are added into the beacon and probe frames, so that the client devices can obtain more information about the surrounding WiFi networks by listening to beacon frames or requesting probe response frames. The key added information elements and the purpose are listed in Table 1.3. Besides, a new query protocol called ANQP is specified for the clients to obtain further information about the AP or the backhaul network services [37]. Some of the important elements are also listed in Table 1.3.

Table 1.3 Important information elements in beacon, probe and ANQP frames in Hotspot 2.0

Parameters	Sub-field	Purpose
Extended capabilities	Interworking	Indicates if this WiFi network can interwork with other networks
	QoS traffic capability	Indicates if the WiFi network can support QoS mapping between WiFi and external networks
Interworking adv.	Access network type	Indicates if the network is private or public, is connected to Internet
	Advertisement protocol ID	Indicate the query protocol ID and the response length limit
Roaming consortium	N/A	Indicate the roaming consortium whose credential can also authenticate with the current AP
ANQP element	WAN metrics	Information about Internet connecting, downlink and uplink speed/load, etc.
	NAI Realm data	Information about NAI realm name and authentication method
	Domain name	Domain information about the operators

1.2.2 Authentication

There are several authentication mechanisms which are applied in different scenarios. Webpage/SMS verification are used in some public places such as airports, malls, etc. However, such mechanism often requires user's manual interaction to input the verification code, which is not feasible for vehicle users, who would prefer the automatic authentication methods. In residential WiFi networks, WPA2-PSK are often used, where a pre-shared credential is stored in WiFi client devices, which will automatically perform the authentication process with the AP. However, the pre-shared credential scheme has some limitations. First, the pre-shared credential is stored at local system of the AP, which lacks of a credential management entity that a vehicle user has to negotiate with the AP when need to update the credentials. WPA2-802.1X is used in commercial WiFi access or enterprises, e.g., eduroam [38, 39], which support remote management via the authentication server. The vehicle users are assigned with a certification to better protect the user credentials. However, the handshake with the remote server requires more management frames to be exchanged and extra backhaul communication delay is introduced. Hotspot 2.0 also employs the 802.1X protocol in the authentication method to provide users secure connection. The latest released version also provisions Online SignUp (OSU) ability which can enable users to automatically select proper plan with reasonable costs from service providers. This kind of management framework makes the WiFi networks as easy and secure as cellular networks. WiFi operators use the authentication step to identify qualified users and prevent unauthorized users from stealing the network resources, while WiFi users rely on this step to protect their communication privacy. There are several authentication mechanisms which are applied in different scenarios. Commonly, webpage/SMS verifications are used in some public places such as airports, malls, etc. And WPA2-PSK are often used in residential areas. WPA2-802.1X is used in commercial WiFi access or enterprises, e.g., eduroam [38]. From the measurement results in [8], it can be observed that the delay caused by the access procedure is mainly from the user authentication step, which takes up the majority part of the management frames and also introduce possible negotiation overhead with remote authentication server. Sophisticated authentication schemes usually require more management frames to be exchanged between the WiFi client and the authentication server, which consumes longer time and thus the overall throughput that the vehicle can achieve would be reduced.

Literature works also investigated efficient authentication methods in vehicular conditions. Chen et al. developed an authentication prototype to support automatic user authentication and seamless interworking between WiFi and WiMAX [40]. Han et al. analyzed the security of connected vehicles and proposed a verification method to protect authenticated users [41]. Fu et al. presented a fast authentication method in heterogeneous WiFi and WiMAX networks to reduce the handover delay [42]. Bohak et al. proposed a fast authentication mechanism to reduce the round trip time between the WiFi user and the remote authentication server [43]. However, such scheme requires to distribute the access information to potential APs, which is difficult in practical deployment. Moustafa et al. applied the 802.11i protocol to setup reliable data transfer in high way condition [44]. These works show the importance to find out the overhead of the authentication step for vehicle users, which can provide guidance for future authentication scheme research and development.

1.2.3 Network Parameters Assignment

To enable effective Internet connection, the vehicle should set proper network parameters, such as valid IP address, correct VLAN configuration, etc. DHCP protocol is often used to dynamically assign the IP address for a WiFi client via a local or remote DHCP server from an address pool. And the Automatic Private IP Address (APIPA) protocol is often used to auto configure the IP address if the DCHP server is not available.

1.2.4 Summary

In this section, we have reviewed the three main steps for a vehicle user to access to a roadside access station. It is inevitable and thus requires carefully scrutiny over the whole access procedure and the impact to the Internet access performance.

1.3 Aim of the Book

In this book, we aim to introduce the Internet access in vehicular networks, considering the practical Internet connection for vehicles via the roadside access points (APs), which normally involve a set of the management frames for control functions. We first propose analytical methods to evaluate the Internet access performance, i.e., the access delay and throughput in Chap. 2. We also consider using the unlicensed spectrum to offload the data traffic from expensive cellular networks to economic IEEE 802.11 IoV utilizing the V2X interworking scheme in Chap. 3. Then we take a deep look at the V2R link management that can adjust the modulation and coding scheme (MCS) smartly to adapt to the highly dynamic channel conditions 4. In addition, in Chap. 5 we employ distributed computing paradigms based on the IoV connectivity for vehicular users, to cooperatively train machine learning (ML) models for intelligent vehicular applications. The conclusion and future directions are given in Chap. 6

References

1. W. Xu, H. Zhou, N. Cheng, F. Lyu, W. Shi, J. Chen, X. Shen, Internet of vehicles in big data era. IEEE/CAA J. Autom. Sin. **5**(1), 19–35 (2018)
2. D. Jiang, L. Delgrossi, IEEE 802.11p: towards an international standard for wireless access in vehicular environments, in *VTC Spring 2008-IEEE Vehicular Technology Conference* (IEEE, Piscataway, 2008), pp. 2036–2040
3. D. Jiang, V. Taliwal, A. Meier, W. Holfelder, R. Herrtwich, Design of 5.9 GHZ DSRC-based vehicular safety communication. IEEE Wirel. Commun. **13**(5), 36–43 (2006)
4. Q. Xu, T. Mak, J. Ko, R. Sengupta, Vehicle-to-vehicle safety messaging in DSRC, in *Proceedings of the 1st ACM International Workshop on Vehicular ad Hoc Networks* (ACM, New York, 2004), pp. 19–28
5. Y.L. Morgan, Notes on DSRC & WAVE standards suite: its architecture, design, and characteristics. IEEE Commun. Surv. Tuts. **12**(4), 504–518 (2010)
6. F. Lyu, H. Zhu, H. Zhou, L. Qian, W. Xu, M. Li, X. Shen, MoMAC: mobility-aware and collision-avoidance MAC for safety applications in VANETs. IEEE Trans. Veh. Technol. **67**(11), 10590–10602 (2018)
7. G. Naik, B. Choudhury, J. Park, IEEE 802.11bd & 5G NR V2X: evolution of radio access technologies for V2X communications. IEEE Access **7**, 70169–70184 (2019)
8. W. Xu, H. A. Omar, W. Zhuang, X. Shen, Delay analysis of in-vehicle internet access via on-road WiFi access points. IEEE Access **5**, 2736–2746 (2017)
9. W. Xu, W. Shi, F. Lyu, H. Zhou, N. Cheng, X. Shen, Throughput analysis of vehicular internet access via roadside WiFi hotspot. IEEE Trans. Veh. Technol. **68**(4), 3980–3991 (2019)
10. N. Cheng, N. Lu, N. Zhang, X. Zhang, X. Shen, J.W. Mark, Opportunistic wifi offloading in vehicular environment: A game-theory approach. IEEE Trans. Intell. Transp. Syst. **17**(7), 1944–1955 (2016)
11. Y. Chen, N. Zhang, Y. Zhang, X. Chen, W. Wu, X.S. Shen, Energy efficient dynamic offloading in mobile edge computing for internet of things. IEEE Trans. Cloud Comput. **9**(3), 1050–1060 (2021)
12. H. Wu, W. Xu, J. Chen, L. Wang, X. Shen, Matching-based content caching in heterogeneous vehicular networks, in *Proc. IEEE Global Communications Conference (GLOBECOM)* (IEEE, Piscataway, 2018), pp. 1–6

13. H. Zhou, N. Cheng, J. Wang, J. Chen, Q. Yu, X. Shen, Toward dynamic link utilization for efficient vehicular edge content distribution. IEEE Trans. Veh. Technol. **68**(9), 8301–8313 (2019)
14. Z. Su, Q. Xu, Y. Hui, M. Wen, S. Guo, A game theoretic approach to parked vehicle assisted content delivery in vehicular ad hoc networks. IEEE Trans. Veh. Technol. **66**(7), 6461–6474 (2017)
15. N. Cheng, N. Lu, N. Zhang, X. Shen, J.W. Mark, Opportunistic wifi offloading in vehicular environment: A queueing analysis, in *IEEE Global Communications Conference (GLOBECOM)* (2014), pp. 211–216
16. H. Viittala, S. Soderi, J. Saloranta, M. Hamalainen, J. Iinatti, An experimental evaluation of WiFi-based vehicle-to-vehicle (V2V) communication in a tunnel, in *IEEE 77th Vehicular Technology Conference (VTC Spring)* (IEEE, Piscataway, 2013), pp. 1–5
17. H.A. Omar, K. Abboud, N. Cheng, K.R. Malekshan, A.T. Gamage, W. Zhuang, A survey on high efficiency wireless local area networks: next generation WiFi. IEEE Commun. Surv. Tutorials **18**(4), 2315–2344 (2016)
18. V.P. Sarvade, S. Kulkarni, Performance analysis of IEEE 802.11 AC for vehicular networks using realistic traffic scenarios, in *IEEE International Conference on Advances in Computing, Communications and Informatics (ICACCI)* (IEEE, Piscataway, 2017), pp. 137–141
19. M.I. Sanchez, A. Boukerche, On IEEE 802.11 K/R/V amendments: do they have a real impact? IEEE Wirel. Commun. **23**(1), 48–55 (2016)
20. W. Xu, H. Zhou, Y. Bi, N. Cheng, X. Shen, L. Thanayankizil, F. Bai, Exploiting hotspot-2.0 for traffic offloading in mobile networks. IEEE Netw. **32**(5), 131–137 (2018)
21. Z. Song, L. Shangguan, K. Jamieson, WiFi goes to town: rapid picocell switching for wireless transit networks, in *Proc. ACM SigCom* (2017), pp. 322–334
22. A.B. Flores, R.E. Guerra, E.W. Knightly, P. Ecclesine, S. Pandey, IEEE 802.11 AF: a standard for TV white space spectrum sharing. IEEE Commun. Mag. **51**(10), 92–100 (2013).
23. H. Zhou, N. Zhang, Y. Bi, Q. Yu, X. Shen, D. Shan, F. Bai, TV white space enabled connected vehicle networks: challenges and solutions. IEEE Netw. **31**(3), 6–13 (2017)
24. H. Zhou, N. Cheng, Q. Yu, X. Shen, D. Shan, F. Bai, Toward multi-radio vehicular data piping for dynamic DSRC/TVWS spectrum sharing. IEEE J. Sel. Areas Commun. **34**(10), 2575–2588 (2016)
25. K. Ishizu, K. Hasegawa, K. Mizutani, H. Sawada, K. Yanagisawa, T. Keat-Beng, T. Matsumura, S. Sasaki, M. Asano, H. Murakami et al., Field experiment of long-distance broadband communications in TV white space using IEEE 802.22 and IEEE 802.11 AF, in *IEEE International Symposium on Wireless Personal Multimedia Communications (WPMC)* (IEEE, Piscataway, 2014), pp. 468–473
26. R. Almesaeed, N.F. Abdullah, A. Doufexi, A.R. Nix, Performance evaluation of 802.11 standards operating in TVWS and higher frequencies under realistic conditions, in *IEEE 80th Vehicular Technology Conference (VTC2014-Fall)* (2014), pp. 1–5
27. TR 21.914, 3GPP TR 21.914 version 14.0.0 Release 14'
28. TR 21.915, 3GPP TR 21.915 version 1.1.0 Release 15
29. 3GPP TR 22.886, Study on enhancement of 3GPP support for 5G V2X services
30. 3GPP TR 37.885, Study on evaluation methodology of new Vehicle-to-Everything V2X use cases for LTE and NR
31. S. Chen, J. Hu, Y. Shi, L. Zhao, LTE-V: A TD-LTE-based V2X solution for future vehicular network. IEEE Internet Things J. **3**(6), 997–1005 (2016)
32. 5G Automotive Association, The case for cellular V2X for safety and cooperative driving, in *5GAA Whitepaper* (2016)
33. S. Vasudevan, K. Papagiannaki, C. Diot, J. Kurose, D. Towsley, Facilitating access point selection in IEEE 802.11 wireless networks, in *Proceedings of the 5th ACM SIGCOMM Conference on Internet Measurement* (2005), pp. 26–26
34. Hotspot 2.0 technical task group, *Wi-Fi Alliance Technical Committee*
35. W. Xu, C. Hua, A. Huang, A game theoretical approach for load balancing user association in 802.11 wireless networks, in *Proc. IEEE Globecom'10* (2010), pp. 1–5

36. Hotspot 2.0 specification and passpoint project. http://www.wi-fi.org/discover-wi-fi/wi-fi-certified-passpoint
37. W. Xu, H. Zhou, Y. Bi, N. Cheng, X. Shen, L. Thanayankizil, F. Bai, Exploiting hotspot-2.0 for traffic offloading in mobile networks. IEEE Netw. **99**, 1–7 (2018)
38. L. Florio, K. Wierenga, Eduroam, providing mobility for roaming users, in *Proceedings of the EUNIS 2005 Conference, Manchester* (2005)
39. Eduroam: secure, world-wide roaming access service for international research and education community. https://www.eduroam.org/
40. Y.-T. Chen, Achieve user authentication and seamless connectivity on WiFi and WiMAX interworked wireless city, in *IFIP International Conference on Wireless and Optical Communications Networks* (2007), pp. 1–5
41. K. Han, S.D. Potluri, K.G. Shin, On authentication in a connected vehicle: secure integration of mobile devices with vehicular networks, in *2013 ACM/IEEE International Conference on Cyber-Physical Systems (ICCPS)* (2013), pp. 160–169
42. A. Fu, G. Zhang, Z. Zhu, Y. Zhang, Fast and secure handover authentication scheme based on ticket for WiMAX and WiFi heterogeneous networks. Wirel. Pers. Commun. **79**(2), 1277–1299 (2014)
43. A. Bohák, L. Buttyán, L. Dóra, An authentication scheme for fast handover between wifi access points, in *Proc. of ACM Wireless Internet Conference (WICON)* (2007)
44. H. Moustafa, G. Bourdon, Y. Gourhant, Providing authentication and access control in vehicular network environment, in *IFIP International Information Security Conference* (Springer, Berlin, 2006), pp. 62–73

Chapter 2
Internet Access Modeling for Vehicular Connection

In this chapter, we focus on the analytical modeling of the Internet access procedure for vehicles. Specifically, the Markov chain model is applied to describe the management frame exchange between the vehicle and the roadside access point. Due to the non-negligible overhead, the access delay is analyzed, which can determine the overall data throughput that can be achieved by the drive-by vehicle. Such access delay and throughput performance is crucial for future IoV network protocol design. We have demonstrated the accuracy of our analysis via both simulation and experimental verification methods.

2.1 Background and Motivation

Due to the ever growing IoV big data, it is expected to support data-rich and bandwidth consuming Internet applications, e.g., in-vehicle infotainment, remote driving, etc. These applications are often deployed onboard to envision immersive experience for both the drivers and the passengers [1, 2]. For example, Intelligent Transportation (ITS) system can collect and disseminate the vehicles' internal and external conditions via the Vehicle-to-Infrastructure (V2I) connections to improve the road efficiency and driving safety level [3, 4]. Besides, with the Internet access, a myriad of infotainment applications, such as video streaming, web page surfing, etc., are becoming indispensable for passengers. Furthermore, some data-craving applications, such as High Definition (HD) map, autonomous driving, etc., is expected to be realized via the high bandwidth connection. It is predicted that the global vehicular data traffic will reach 300 Zettabytes by 2020 [5], which can cause a great pressure to current Internet access technologies for vehicles [6]. Cellular networks are initially adopted for Internet access for vehicles. However, the costs of downloading/uploading all traffic to cellular networks are usually not affordable for vehicle users. In addition, cellular network capacity will be drained

W. Xu et al., *Internet Access in Vehicular Networks*,
https://doi.org/10.1007/978-3-030-88991-3_2

up in dense condition where lots of vehicles are requesting heavy data tasks [7]. To overcome the drawbacks of cellular Internet access, different wireless technologies have been proposed to provide alternate choices. Zhou et al. used the TV white space (TVWS) spectrum enabled infostation to disseminate the multimedia content [8]. However, the adoption of TVWS is restricted by the geo-location where the regulation and policy of using TVWS spectrum varies place by place. Wu et al. adopted the heterogeneous small cells to offload the cellular traffic, however, it requires frequent vertical and horizontal handoff for vehicle users due to their high mobility [9]. Ligo et al. utilized the Dedicated Short Range Communications (DSRC) to offload the vehicular traffic to Internet [10]. However, the link rate is limited as the DSRC bandwidth is only half of 802.11a. Luo et al. investigated the inter-vehicle performance based on the visible light communication (VLC) [11], which is greatly affected by the day light noise and line-of-sight condition and the network deployment is difficult.

WiFi can overcome the restriction of the above radio technologies. First, WiFi has been widely used for Internet access around the world for years. It is predicted that by the year of 2021, 73% of the global Internet traffic will be served by WiFi networks [12]. WiFi devices are universally compatible and the unlicensed spectrum is used that are not restricted over all regions of the world. Compared with licensed spectrum based technologies, e.g., LTE-V2X, WiFi device does not need a permission, and different generation WiFi can communicate with each other without upgrading their devices. Secondly, WiFi has significant link throughput. The latest 802.11ac protocol can achieve the peak link rate around 1 Gbps [13], which provides enough capacity for vehicles in a WiFi cell even in dense condition. Furthermore, unlike deploying TVWS station or consuming in cellular networks, the economic cost of operating WiFi networks are relatively low. A roadside WiFi network can be agilely setup using commercial off-the-shelf devices and open source software, which is much cheaper than building infrastructures for macro-cells, e.g., LTE-V2X base stations [14].

There have been extensive research works on roadside WiFi Internet access for vehicles. Ott et al. first proposed the concept of 'Drive-thru Internet' that utilize a 802.11b hotspot to provide temporal Internet access for drive-by vehicles [15]. The conducted road test showed that considerable data traffic can be transmitted between the roadside hotspot and the vehicle, which demonstrated that it is feasible to provide Internet access for vehicles by WiFi technologies. Mahajan et al. measured the end-to-end connectivity between moving vehicles and the roadside WiFi Access Points (APs). And the throughput performance between the AP and the moving vehicles are also investigated on different regions [16]. Cheng et al. adopted the queueing model to analyze the traffic offloading performance for vehicles using the intermittent roadside WiFi networks. The relationship between the offloading effectiveness and the average service delay of the data tasks are analyzed [17]. Similarly, Zhou et al. proposed a cluster based scheme to conduct cooperation between multiple roadside APs to deliver content to vehicles [18]. However, existing works seldom consider the access procedure that a vehicle user has to accomplish before he/she can actually access to Internet via a roadside hotspot [19]. The access procedure generally contains the three steps as mention in Sect. 1.2.

(1) Network Detection: Before access to the roadside WiFi network, the vehicle should perceive the existence of the network via beacon frames exchange or other query protocols (e.g., the Access Network Query Protocol (ANQP) of Hotspot 2.0) [20].
(2) User Authentication: It is required to setup reliable and secure wireless connections between the AP and its users.
(3) Network Parameters Assignment: To communicate with other entities on Internet, it is required that the associated WiFi users to have an IP address. Dynamic Host Configuration Protocol (DHCP) protocol is often used to assign the IP address for WiFi users dynamically. In some conditions, the vehicle might need to configure extra network settings.[1] Such steps require a number of management frames to be exchanged between the vehicle, hotspot and remote servers.

In practice, the above steps are necessary to setup effective and reliable Internet connection for vehicles. However, most previous research works and conducted experiments have neglected these steps. In [15], Ott et al. used an open WiFi hotspot without verifying user credential before allowing network access, i.e., no user authentication step is included. And a static IP address was assigned to the vehicle who could access to the network immediately when the vehicle drives over. The experiment in [21] did not include the authentication step, while the measurement result showed that the DHCP latency could be several seconds. Mahajan et al. also did not consider the overhead of authentication and IP address acquisition in their measurements in [16]. The analytical works from [17, 18, 22, 23] assume that Internet can be accessed as soon as the vehicle drives into the cell coverage areas, where the impacts of the access procedure were omitted.

A vehicle cannot access to Internet via the roadside WiFi AP until the access procedure is accomplished. Since the sojourn time of a vehicle within the WiFi coverage area is limited, a fast access procedure will leave more time for downloading/uploading Internet data, and vice versa. The accomplishment of the access procedure can be deferred in the following conditions. First, when the number of other WiFi clients, e.g., other vehicles, are associated to the same AP increases, the finish of the procedure will be deferred as the exchange of all the management frames requires more time when contending channel resource with its peers. Secondly, when the channel quality degrades, each management frame may require more re-transmission attempts as the packet error rate increase. Besides, the parameters of the IEEE 802.11 distributed coordination function (DCF), e.g., minimum window size, back off stage number, may also change the DCF process for transmitting each management frame [24], which lead to different consequence of a transmission attempt, and thus affect the accomplishment of the access procedure.

[1] For example, apply VLAN settings to divide the broadcast domain of several sub-nets.

Furthermore, different authentication methods require the vehicle to exchange different set of management frames with the AP and authentication server, and thus the steps of user authentication are different.

In this chapter, we investigate the performance of drive-thru Internet considering the accomplishment of the practical access procedure. Particularly, our objective is to find the relationship between the access delay and throughput via a roadside AP, and environmental and protocol execution conditions such as the channel conditions, contention level, network protocol configurations, etc. by considering the overhead of the Internet access procedure. The overhead of network detection via query or beaconing, and network parameters assignment were not considered in status quo literature. Shin et al. utilized a selective channel scanning method to shorten the network detection delay and thus the access overhead can be reduced [25]. However, the authentication step which takes the majority part of the access delay was not considered. In fact, to eliminate the impacts of the access procedure, which are difficult to evaluate, most experiment and measurement of the WiFi connection between vehicles and APs in literature works applied the open association [15, 26], static IP address setting [16, 27] and without authentication step [21, 28]. Lu et al. argued that the time for access procedure cannot be neglected due to high mobility of vehicles, which can take up to ten or more seconds [1]. Our purpose is to analyze that in a practical scenario where the three steps of the access procedure cannot be waived, what is the Internet access performance of the IoV connection.

2.2 Delay Analysis of Vehicular Internet Access

Limited existing works focus the time duration that a vehicle user needs to take before the user can access the service of an on-road WiFi AP and actually connect to the Internet [29]. This time duration, referred to as 'access delay', is required mainly to perform the authentication and Internet Protocol (IP) address assignment. In [15], the conducted drive-thru experiments employ an open access scheme and a static IP address, which allow a vehicle user to automatically associate and access the AP service, without any consideration of the access delay. Yet, except for experimental testing or research purposes, the access delay is unavoidable to perform the authentication procedure, which is essential for WiFi network users and operators.

We consider the WPA2 and Hotspot 2.0 authentication methods that can be automatically accessed without any manual interaction. In addition to the time duration required to complete the authentication procedure, another duration is needed for a vehicle user to obtain an IP address, e.g., via Dynamic Host Configuration Protocol (DHCP) protocol. The sum of the durations required for authentication and IP assignment constitutes the access delay, which can last for a few seconds [21]. In such a case, a vehicle user can have a limited time to utilize the Internet resources before the vehicle moves out of the coverage area of a WiFi AP, especially with a high vehicle moving speed. Hence, a 'quickWiFi' scheme was proposed to reduce the access delay by tuning related WiFi parameters and optimizing the AP scanning strategy for clients [26].

The access delay can be affected in several ways. First, if the AP is serving a large number of users, the access delay will increase for a new user due to a high level of channel contention using the IEEE 802.11 standard distributed coordination function (DCF). Second, a poor wireless propagation channel can result in a high frame error rate, which further increases the access delay, due to re-transmission of management frames that are not successfully delivered. Third, different authentication protocols require different sequences of management frame exchanges between the AP and a new user, leading to a different access delay associated with each authentication methods. To the best of our knowledge, the effects of the number of contending WiFi users, the wireless channel conditions, and the employed authentication method on the access delay have not been analyzed. We investigate how these factors affect the access delay. We propose a Markov chain-based analytical model that can be applied for any authentication method, in order to calculate the average access delay, given the time-varying channel conditions and number of contending WiFi users in a vehicular environment. The accuracy of the proposed analytical model is studied via MATLAB simulations and experimental testing. The experimental testing is conducted using commercial off-the-shelf (COTS) WiFi products supporting the IEEE 802.11n standard, together with an advanced channel emulator that emulates the wireless channel conditions between the vehicles and a WiFi AP in an expressway scenario. The analytical, simulation, and experimental testing results of the average access delay are obtained for the WPA2-PSK and WPA2-802.1X authentication methods, under various wireless channel conditions and for various numbers of contending WiFi users.

2.2.1 System Model

We consider a single WiFi AP that provides Internet connectivity for vehicles on the road. When a vehicle enters the communication range of the AP, before connecting to the Internet, the vehicle exchanges a sequence of management frames with the AP in order to perform the necessary procedures for authentication and IP address allocation. The management frame exchanges between the vehicle and the AP depend on the WiFi network access standard, e.g., WPA2 [30] and Hotspot 2.0 [31], and the authentication mechanism, e.g., IEEE 802.1X [32] and extensible authentication protocol (EAP) [33, 34].

For instance, Figs. 2.1 and 2.2 respectively show the sequence of management frames exchanged between a vehicle and the AP for the WPA2-PSK and WPA2-802.1X authentication methods. The generation of some management frames may require communication between the AP and a remote server through a core network. For instance, as shown in Fig. 2.2, the AP needs to connect to a remote authentication, authorization, and accounting (AAA) server before replying to some

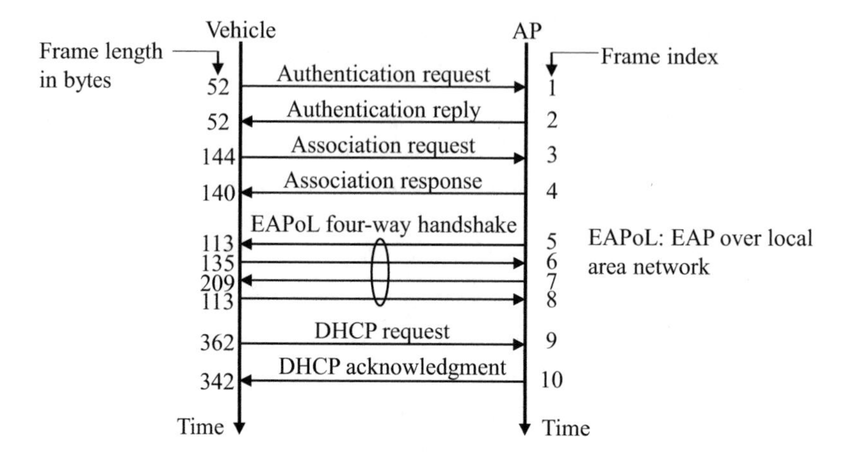

Fig. 2.1 Management frames exchanged between a vehicle and an AP based on the WPA2-PSK mode for authentication ($N_f = 10$)

frames from a vehicle. We focus on a single vehicle, referred to as tagged vehicle, that just enters the communication range of the AP and attempts to connect to the Internet via the AP. To perform this Internet connection, the management frames exchanged between the tagged vehicle and the AP, as shown in Figs. 2.1 and 2.2, are indexed from 1 to N_f, and the length of the ith management frame is denoted by $l_i, i = 1, .., N_f$. The frame length indicates the length of the data field of the physical layer (PHY) protocol data unit (PPDU), which consists of the encoded MAC layer protocol data unit (MPDU) and other fields that are included by the PHY and transmitted over-the-air using the same bit rate as the MPDU, such as the service field and tail bits added by the IEEE 802.11 orthogonal frequency division multiplexing (OFDM) PHY standard [35].

In addition to the tagged vehicle, there exist a number of neighbor vehicles that are already connected to the Internet via the AP and uploading data to the AP. It is assumed that, each neighbor vehicle always has a data frame to upload to the AP, from the instant that the tagged vehicle enters the communication range of the AP until all the N_f management frames are successfully exchanged. Each data frame uploaded by a neighbor vehicle has a fixed length denoted by l, and is transmitted at a constant PHY bit rate denoted by r. All the nodes (i.e., the neighbor vehicles, the tagged vehicle, and the AP) are within the communication range of each other and employ the IEEE 802.11 DCF to access the channel [35], with a minimum contention window size denoted by w, and a number of back-off stages indexed from 0 to $m - 1$, where m denotes the total number of back-off stages in the absence of request-to-send/clear-to-send (RTS/CTS) handshaking. At each back-off stage, the tagged vehicle and the AP employs a PHY bit rate, denoted by $r_{ib}, i = 1, \ldots, N_f$ and $b = 0, \ldots, m - 1$, for the next transmission attempt of the ith management frame that is being exchanged. For the same management frame index, i, the values of r_{ib} $\forall b$ are determined based on a certain rate switching

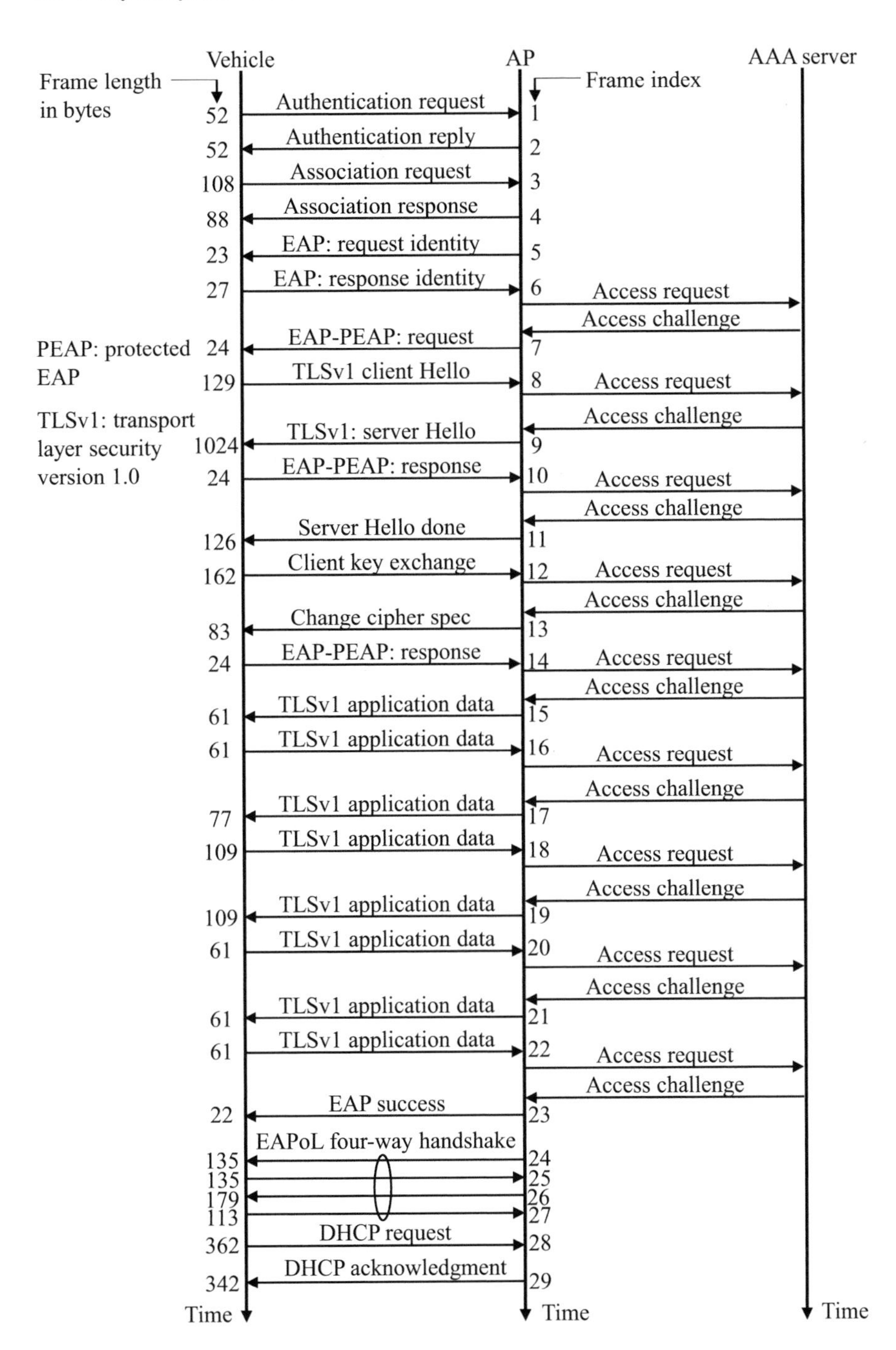

Fig. 2.2 Management frames exchanged between a vehicle and an AP based on the WPA2-802.1X mode for authentication ($N_f = 29$)

algorithm, while for the same back-off stage index, b, the value of r_{ib} depends on whether the tagged vehicle or the AP is the source of the ith management frame. If a management/data frame is successfully received, an acknowledgment (ACK) frame of length a is transmitted using the same PHY bit rate as that for the management/data frame transmission. On the contrary, if a management/data frame is not successfully delivered to its destination, the frame is referred to as a 'lost' frame. The ACK timeout duration that the source of a lost frame needs to wait for, before invoking the DCF back-off procedure, is neglected [35]. A lost frame is retransmitted by its source node until it is successfully delivered, without any maximum retry limit.

When the tagged vehicle or the AP attempts to transmit the ith management frame, $i = 1, \ldots, N_f$, the total number of nodes that are contending to access the channel is constant and denoted by n_i, which consists of all the neighbor vehicles plus one node (i.e., either the AP or the tagged vehicle, depending on which one is the source of the ith management frame). For the n_i contending nodes, $i = 1, \ldots, N_f$, let τ_i denote the probability that a node transmits a frame in a randomly selected slot duration,[2] α_i the probability that a transmitted frame is lost due to a transmission collision, β_i the probability that a transmitted frame is lost due to a poor channel condition ($0 < \beta_i < 1$), and δ_i the probability that a transmitted frame is lost due to a transmission collision or poor channel, i.e., $\delta_i = 1 - (1 - \alpha_i)(1 - \beta_i)$. It is assumed that the value of each of α_i, β_i, and (consequently) δ_i, $i = 1, \ldots, N_f$, is the same for any frame transmitted by any of the n_i contending nodes, and remains constant until the ith management frame is successfully exchanged between the tagged vehicle and the AP. Also, the success events of different delivery trials of the same management/data frame are independent. If a transmission collision happens among management and data frames, none of the contending nodes can successfully receive any of the colliding frames. On the contrary, if no transmission collision happens for a transmitted frame, but the frame is lost due to a poor channel condition, the back-off procedure of each node that successfully received the frame is invoked immediately at the end of transmission of the lost frame, i.e., the additional wait time that consists of short interframe space (SIFS) and ACK transmission durations is neglected [35].

In the following, the notation $\mathbb{E}(Y)$ denotes the expected value of a random variable Y, $\mathbb{E}(Y|Z = z)$ the conditional expected value of Y given the event that another random variable Z takes the value z, and $\max(a, b)$ the maximum of the two values a and b.

[2] The slot duration is defined as the duration between two consecutive variations in the back-off counter or back-off stage of a contending node [36].

2.2.2 Access Delay Analysis

The objective of this section is to derive the average access delay that is required for the tagged vehicle and the AP to complete the authentication and IP allocation procedures by exchanging the necessary N_f management frames. First, we define a time step as the sum of the durations required by the source of a management frame to:

(a) generate the frame,
(b) complete the DCF back-off procedure and start the over-the-air transmission of the frame, and
(c) either successfully transmit the frame and receive the corresponding ACK frame or unsuccessfully transmit the frame and wait until the channel is sensed idle (the earlier of the two events).

Based on the definition, the access delay from the time instant that the first management frame is being generated until all the N_f management frames are successfully exchanged can be partitioned into a sequence of time steps. At the start of each time step, a management frame is required to be (re)transmitted either by the tagged vehicle or by the AP. Let X_n be the index of the management frame that should be exchanged between the tagged vehicle and the AP at the start of the nth time step. Based on the system model, X_n is a discrete-time Markov chain that takes integer values from 1 to N_f. Additionally, the value of $N_f + 1$ is added to the state space of X_n to represent the event that all the N_f frames are successfully exchanged between the tagged vehicle and the AP.[3] Hence, when $X_n = i, i = 1, \ldots, N_f$, the Markov chain either transits to state $i + 1$ or remains at its current state, based on whether or not the transmission of the ith frame is successful at the end of the nth time step, as illustrated in Fig. 2.3. Therefore, in order to calculate the average access delay, the main idea is to find the average duration that the Markov chain X_n needs in order to transit from state 1 to state $N_f + 1$ for the first time. The remainder of this section shows how this average duration can be obtained.

For Markov chain X_n, let p_{ij} denote the one-step transition probability from state i to state j, where

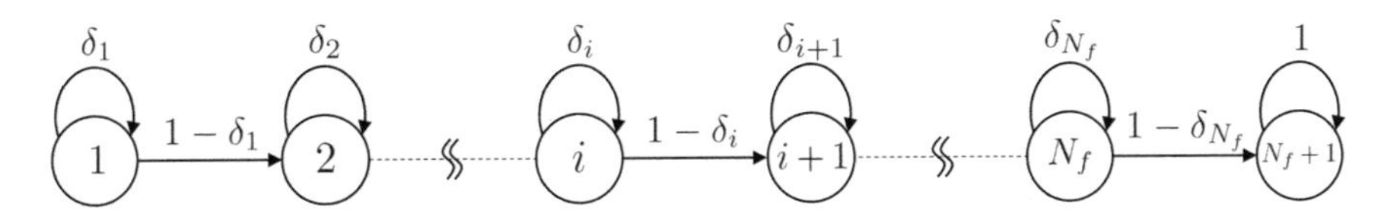

Fig. 2.3 Illustration of the Markov chain and one-step transition probabilities for states 1 to N_f+1

[3] When $X_n = N_f + 1$, the kth time step, $k \geq n$, can take any positive value.

$$p_{ij} = \begin{cases} \delta_i, & i = j = 1, \ldots, N_f \\ 1, & i = j = N_f + 1 \\ 1 - \delta_i, & i = j - 1 = 1, \ldots, N_f \\ 0, & \text{elsewhere.} \end{cases} \tag{2.1}$$

In (2.1), the value of δ_i can be obtained by extending Bianchi's DCF model [36] to account for the frame loss due to channel conditions.[4] That is, for each $i = 1, \ldots, N_f$, the value of δ_i is calculated by solving the system of Eqs. (4.1a)–(2.19c) in variables τ_i, α_i, and δ_i:

$$\tau_i = \frac{2(1 - 2\delta_i)}{(1 - 2\delta_i)(w + 1) + \delta_i w(1 - (2\delta_i)^{m-1})} \tag{2.2a}$$

$$\alpha_i = 1 - (1 - \tau_i)^{n_i - 1} \tag{2.2b}$$

$$\delta_i = 1 - (1 - \alpha_i)(1 - \beta_i). \tag{2.2c}$$

To show that there exists a unique value for each of τ_i, α_i, and δ_i, from (2.19c) and (2.19b) we have

$$\tau_i = 1 - \left(\frac{1 - \delta_i}{1 - \beta_i}\right)^{\frac{1}{n_i - 1}}. \tag{2.3}$$

Therefore, using (2.19a) and (2.3), we can prove the existence and uniqueness of the solution for the system of the three equations (2.19a)–(2.19c) following a similar approach as in [36]. Given the one-step transition probabilities in (2.1), the first passage time probabilities can be obtained using

$$f_{ij}^{(1)} = p_{ij} \tag{2.4a}$$

$$f_{ij}^{(n)} = \sum_{\substack{k=1 \\ k \neq j}}^{N_f + 1} p_{ik} f_{kj}^{(n-1)}, \quad n > 1 \tag{2.4b}$$

where $f_{ij}^{(n)}$ denotes the n-step first passage time probability from state i to state j. Note that, for the Markov chain in Fig. 2.3, $\sum_{n=1}^{\infty} f_{ij}^{(n)} = 1$ iff $j > i$ or $j = i = N_f + 1$, provided that $\delta_i \neq 1\ \forall i$. Now, let D_{ij} denote the first passage delay from state i to state j, i.e., the delay that the Markov chain requires to transit to state

[4] When the tagged vehicle or the AP attempts to transmit the ith management frame, $i = 1, \ldots, N_f$, each of the n_i contending nodes always has a frame to transmit, i.e., in a traffic saturation conditions [36], until the ith frame is successfully delivered.

j for the first time, given that the Markov chain is currently at state i, where $i = 1, \ldots, N_f$, $j = 1, \ldots, N_f + 1$, and $j > i$. By using the law of total expectation and the first passage time probabilities from (2.4a)–(2.4b), and by noting that $f_{ij}^{(n)} \neq 0$ only if $n \geq j - i$ (Fig. 2.3), the expected value of D_{ij} is given by

$$\mathbb{E}(D_{ij}) = \sum_{n=j-i}^{\infty} \mathbb{E}(D_{ij}^{(n)}) f_{ij}^{(n)},$$
$$i, j \in \{1, \ldots, N_f + 1\} \text{ and } i < j \tag{2.5}$$

where $D_{ij}^{(n)}$ denotes the n-step first passage delay from state i to state j, i.e., the delay that the Markov chain requires to transit to state j for the first time in n time steps, given that the Markov chain is currently at state i. Consequently, the average access delay can be directly obtained from (2.5), by setting $i = 1$ and $j = N_f + 1$. However, in order to evaluate (2.5) for specific i and j values, the expected value $\mathbb{E}(D_{ij}^{(n)})$ should be calculated $\forall n \in \mathbb{N}^+$ such that $n \geq j - i$. For $n \geq j - i$ and $n \neq 1$, the value of $\mathbb{E}(D_{ij}^{(n)})$ can be obtained in a recursive way as follows. Let random variable $K_{ij}^{(n)}$ denote the index of the first state to which the Markov chain transits from state i, given that the Markov chain transits from state i to state j for the first time in n steps, where $i, j = 1, \ldots, N_f, i < j$, and $n \geq \max(j - i, 2)$. For these i, j, and n values, let set $\Omega_{ij}^{(n)} = \{k : p_{ik} \neq 0 \text{ and } j - n + 1 \leq k < j\}$ denote all possible values of random variable $K_{ij}^{(n)}$, which is given by

$$\Omega_{ij}^{(n)} = \begin{cases} \{i\}, & j = i + 1 \\ \{i + 1\}, & j = i + n \\ \{i, i + 1\}, & \text{elsewhere.} \end{cases} \tag{2.6}$$

Hence, the expected value $\mathbb{E}(D_{ij}^{(n)})$ can be calculated by using

$$\begin{aligned} \mathbb{E}(D_{ij}^{(n)}) &= \sum_{k \in \Omega_{ij}^{(n)}} \mathbb{E}(D_{ij}^{(n)} | K_{ij}^{(n)} = k) \frac{p_{ik} f_{kj}^{(n-1)}}{f_{ij}^{(n)}} \\ &= \sum_{k \in \Omega_{ij}^{(n)}} \left(\mathbb{E}(D_{ik}^{(1)}) + \mathbb{E}(D_{kj}^{(n-1)}) \right) \frac{p_{ik} f_{kj}^{(n-1)}}{f_{ij}^{(n)}}, \end{aligned} \tag{2.7}$$
$$i, j \in \{1, \ldots, N_f + 1\}, i < j, \text{ and } n \geq \max(j - i, 2).$$

In order to evaluate $\mathbb{E}(D_{ij}^{(n)})$, it is required to find the values of $\mathbb{E}(D_{ik}^{(1)})$, $\forall i \in \{1, \ldots, N_f\}$ and $k \in \{i, i + 1\}$. First, we have

$$\begin{aligned}\mathbb{E}(D_{ik}^{(1)}) &= \mathbb{E}(U_i) + \mathbb{E}(V_i) + \text{DIFS} + \mathbb{E}(R_{ik}),\\ &i \in \{1, \ldots, N_f\} \text{ and } k \in \{i, i+1\}\end{aligned} \tag{2.8}$$

where U_i is the processing time at the start of a time step required to generate the ith management frame, including the duration needed for communication through the core network (if exists); V_i is the time spent until the channel is sensed idle and the back-off procedure is invoked by the source of the ith management frame; DIFS is the duration of a DCF interframe space [35]; and R_{ik} is the remainder of a time step, excluding the U_i, V_i, and DIFS durations, when the ith management frame is either successfully ($k = i + 1$) or unsuccessfully ($k = i$) delivered. The processing time, U_i, of the ith management frame is nonzero only before the first transmission attempt of the frame (i.e., when the source of the frame is at back-off stage 0). When $U_i = 0$, we have $V_i = 0$ in consequence, since each time step starts at a moment the channel already starts to become idle.[5] In order to calculate $\mathbb{E}(U_i)$, $\mathbb{E}(V_i)$, and $\mathbb{E}(R_{ik})$, $k \in \{i, i + 1\}$, for a specific value of $i \in \{1, \ldots, N_f\}$, let random variable B_i denote the back-off stage of the source node that attempts to transmit the ith management frame at the start of a time step. The probability distribution function of B_i is given by

$$P_{B_i}(b) = \begin{cases} \delta_i^b(1-\delta_i), & b = 0, \ldots, m-2 \\ 1 - \sum\limits_{q=0}^{m-2} \delta_i^q(1-\delta_i), & b = m-1. \end{cases} \tag{2.9}$$

Hence,

$$\begin{aligned}\mathbb{E}(U_i) &= \sum_{b=0}^{m-1} \mathbb{E}(U_i|B_i = b)P_{B_i}(b)\\ &= \mathbb{E}(U_i|B_i = 0)P_{B_i}(0)\end{aligned} \tag{2.10}$$

$$\begin{aligned}\mathbb{E}(V_i) &= \sum_{b=0}^{m-1} \mathbb{E}(V_i|B_i = b)P_{B_i}(b)\\ &= \mathbb{E}(V_i|B_i = 0)P_{B_i}(0)\end{aligned} \tag{2.11}$$

$$\begin{aligned}\mathbb{E}(R_{ik}) &= \sum_{b=0}^{m-1} \mathbb{E}(R_{ik}|B_i = b)P_{B_i}(b),\\ &i \in \{1, \ldots, N_f\} \text{ and } k \in \{i, i+1\}.\end{aligned} \tag{2.12}$$

[5] An exception is the first time step when $i = 1$, for which the value of V_1 is neglected.

In (2.10), the value of $\mathbb{E}(U_i|B_i = 0)$ can be found for a given probability density function of U_i, while in (2.11), the value of $\mathbb{E}(V_i|B_i = 0)$ can be approximated as the duration of a successful over-the-air delivery of a data frame, i.e.,

$$\mathbb{E}(V_i|B_i = 0) = h + \frac{l}{r} + \text{SIFS} + \frac{a}{r} \tag{2.13}$$

where h is the transmission duration of PHY information other than the PPDU data field, e.g., PHY convergence procedure (PLCP) preamble and signal fields of the IEEE 802.11 OFDM PHY [35]. In (2.12), the conditional expectation $\mathbb{E}(R_{ik}|B_i = b)$ can be calculated using (2.14a)–(2.14b) as follows:

$$\mathbb{E}(R_{ii+1}|B_i = b) = \mathbb{E}(C_b)\mathbb{E}(S_i) + y_{ib} \tag{2.14a}$$

$$\begin{gathered}\mathbb{E}(R_{ii}|B_i = b) = \mathbb{E}(C_b)\mathbb{E}(S_i) + z_{ib} \\ i \in \{1, \ldots, N_f\} \text{ and } b \in \{0, \ldots, m-1\}\end{gathered} \tag{2.14b}$$

where C_b denotes the value of the back-off counter of the source node at back-off stage b, S_i the duration required to decrease the back-off counter of the source node by 1 when attempting to transmit the ith management frame, and y_{ib} (z_{ib}) the remainder of a time step after the over-the-air transmission of the ith management frame starts, when the transmission is successful (unsuccessful) and the source node is at the bth back-off stage. Since at the bth back-off stage, the value of the back-off counter is equally likely selected from 0 to $2^b w - 1$ [35], the expected value $\mathbb{E}(C_b)$ is given by

$$\mathbb{E}(C_b) = \frac{2^b w - 1}{2}, b \in \{0, \ldots, m-1\}. \tag{2.15}$$

The values of y_{ib} and z_{ib} can be calculated (by neglecting the propagation delay) using

$$y_{ib} = h + \frac{l_i}{r_{ib}} + \text{SIFS} + \frac{a}{r_{ib}} \tag{2.16a}$$

$$z_{ib} = h + \frac{\beta_i(1-\alpha_i)}{\delta_i}\frac{l_i}{r_{ib}} + \frac{\alpha_i}{\delta_i}\max(\frac{l_i}{r_{ib}}, \frac{l}{r}) \tag{2.16b}$$

$$i \in \{1, \ldots, N_f\} \text{ and } b \in \{0, \ldots, m-1\}.$$

Note that, in (2.16b), the values of $\frac{\beta_i(1-\alpha_i)}{\delta_i}$ and $\frac{\alpha_i}{\delta_i}$ respectively equal the probability that a failure of delivering the ith management frame is due to a poor channel condition only (i.e., no transmission collision) or involves a transmission collision with a data frame. Finally, the value of $\mathbb{E}(S_i)$ can be obtained using (2.17a)–(2.17c), given by

$$\mathbb{E}(S_i) = (1 - \zeta_i)\sigma + \zeta_i\left(h + \frac{l}{r} + \text{DIFS}\right) + \nu_i\left(\text{SIFS} + \frac{a}{r}\right) \tag{2.17a}$$

$$\zeta_i = 1 - (1 - \tau_i)^{n_i - 1} \tag{2.17b}$$

$$\nu_i = (1 - \beta_i)(n_i - 1)\tau_i(1 - \tau_i)^{n_i - 2} \tag{2.17c}$$

$$i \in \{1, \ldots, N_f\}$$

where σ is the idle slot duration, ζ_i and ν_i respectively denote the probability of a transmission and the probability of a successful transmission in a slot duration from the $n_i - 1$ nodes that are contending with the source node of the ith management frame. By using (2.1)–(2.4) and (2.6)–(2.17), the expected value of the first passage delay, $\mathbb{E}(D_{ij})$, from a state i to another state j can be obtained from (2.5). By setting $i = 1$ and $j = N_f + 1$, the value of $\mathbb{E}(D_{1N_f+1})$ represents the average access delay.

2.2.3 Delay Analysis and Simulation

This section provides numerical results based on the mathematical analysis in Sect. 2.2.2 to investigate the access delay performance with respect to the number of contending nodes, the wireless channel conditions, and the associated authentication mechanisms. The first authentication mechanism under consideration is based on the WPA2-802.1X mode, which is used for enterprise networks and requires an authentication server [32]; while the second authentication mechanism is based on the WPA2-PSK mode, which is mainly employed for home and small office networks and does not require an authentication server [33]. The two authentication mechanisms result in two different sequences of management frame exchanges between the AP and the tagged vehicle, as well as different values of the additional delay introduced for some management frames due to possible communication between the AP and an authentication server through the core network. The numerical results are generated based on the IEEE 802.11n standard, which (together with the authentication mechanism) defines the sequence of management frames that should be exchanged for the tagged vehicle to connect to the Internet through the AP. When delivering the management frames, the values of each of β_i and n_i $\forall i$ (Sect. 2.2.2) are set to fixed values, denoted by β and n, respectively. Similarly, for the ith management frame, the values of r_{ib} $\forall b$ are set to a fixed value, denoted by $r_i, i = 1, \ldots, N_f$, where each r_i is set to the bit rate employed by the source of the ith management frame at back-off stage 0, as obtained from the experimental testing in Sect. 2.2.4. The experiment in Sect. 2.2.4 also provides the average processing delay for each management frame, $\mathbb{E}(U_i|B_i = 0)$ $\forall i$ in (2.10). This section also includes computer simulations using MATLAB, in order to study the accuracy of the mathematical analysis presented in Sect. 2.2.2. We simulate the exchange of the N_f management frames between the tagged vehicle and the AP, for the WPA2-PSK and WPA2-802.1X authentication modes, based on the IEEE 802.11 DCF for channel access by all nodes. For each combination of n and β values in the simulations, the average access delay required to exchange the N_f frames is estimated by using 200 samples (i.e., 200 repetitions of successful delivery of all the N_f frames), which result in acceptable 95% confidence interval for all the n and β values under consideration for each authentication mode. The parameter values used to obtain the analytical, simulation, and experimental results are summarized in Table 2.1.

Figure 2.4 shows the access delay performance when the WPA2-802.1X mode is used for authentication. As shown in Fig. 2.4a, the average access delay increases almost linearly with the number of contending nodes, n, for a given wireless channel represented by the probability, β, that a frame is lost due to a poor channel condition. The rate of average access delay increase with n is higher when the β value increases. For instance, in Fig. 2.4, the rate of increase of the curve corresponding to $\beta = 0.6$ is approximately double that of the curve corresponding to $\beta = 0.1$. The effect of β on the average access delay is illustrated in Fig. 2.4b for different n values. When the value of n is small ($n \leq 5$), increasing β up to 0.5 does not result in a significant increase in the average access delay. The reason is that, if a management frame is lost due to channel conditions, the additional delay required to regain access of the channel and retransmit the fame is not significant when n is small, due to a low channel contention level. On the contrary, when the n value increases, the effect of β on the average access delay becomes more noticeable, as shown in Fig. 2.4b. When β approaches 1, the average access delay tends to ∞, as expected, since no management frame can be successfully delivered. There is a good match between the analytical and simulation results, which indicates the accuracy of the analytical model presented in Sect. 2.2.2. The same behavior of the average access delay illustrated in Fig. 2.4 for the WPA2-802.1X standard is observed for larger n values (up to 150) and when the WPA2-PSK mode is used for authentication. However, when WPA2-PSK is employed, the average access delay is considerably lower than that of the WPA2-802.1X mode, due to a smaller number of management frames required to achieve the Internet access (Figs. 2.1 and 2.2). Figure 2.5 compares the average access delay for the WPA2-802.1X and the WPA2-PSK modes for different n and β values. The average access delay and its rate of increase with respect to n are higher for the WPA2-802.1X mode as compared with the WPA2-PSK for all the n and β values shown. Results in this section help to understand the behavior of the average access delay under various channel conditions, with different number of contending nodes, and using the different authentication methods, which is useful to select or develop a suitable WiFi network access scheme for a vehicular environment.

Table 2.1 Parameter values used to generate the analytical, simulation, and experimental results

Param	Value	Param	Value	Param	Value	Param	Value
w	16	DIFS	SIFS + 2σ	N_f for WPA2-802.1X	29 frames	a	32 bytes
m	7	Preamble length	16 μs	N_f for WPA2-PSK	10 frames	l	1574 bytes
σ	9 μs	PLCP header length	4 μs	r_i (ith frame transmitted by the AP)	24 Mbps	l_i for WPA2-802.1X	Fig. 2.2
SIFS	16 μs	h	Preamble length + PLCP header length	r_i (ith frame transmitted by the tagged vehicle)	6 Mbps	l_i for WPA2-PSK	Fig. 2.1
r	24 Mbps	$\mathbb{E}(U_i)$ for WPA2-802.1X	Varies from 45 μs to 64 ms for $i = 1, \ldots, N_f$	$\mathbb{E}(U_i)$ for WPA2-PSK	Varies from 87 μs to 70 ms for $i = 1, \ldots, N_f$	–	–

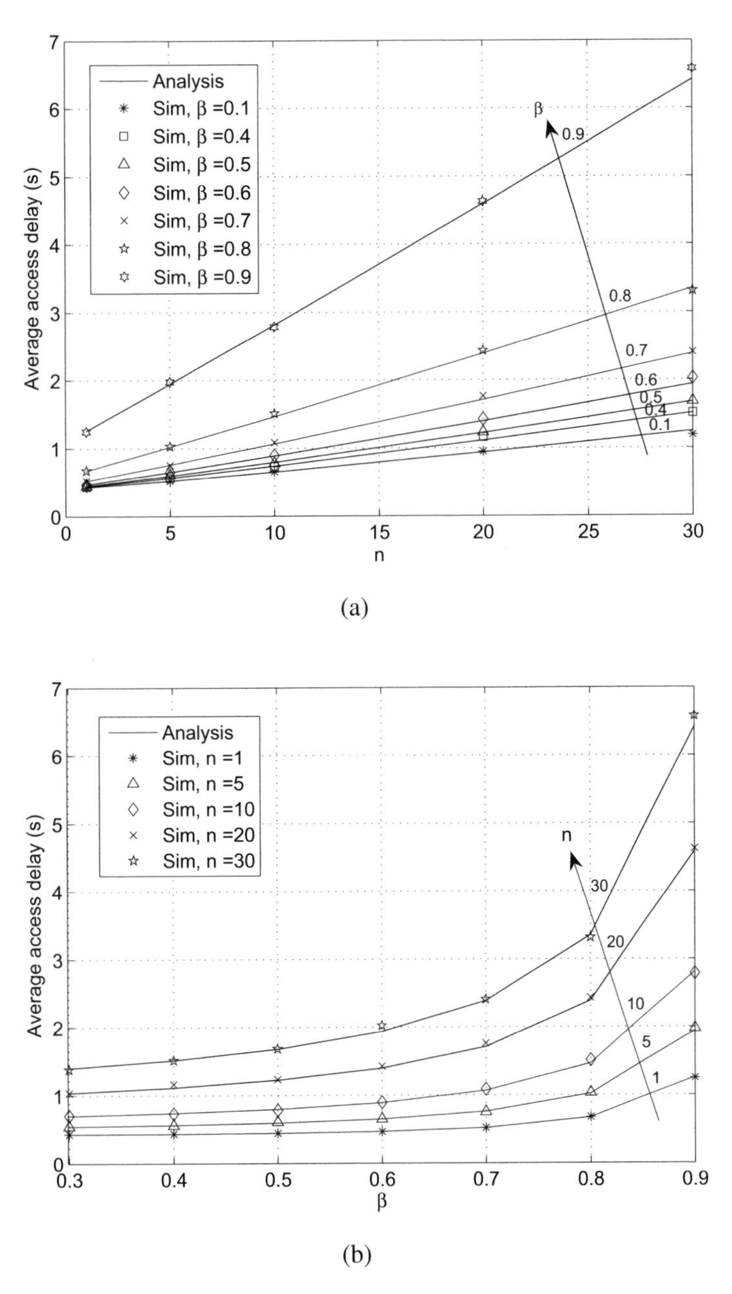

Fig. 2.4 Analytical and simulations (Sim) results of the access delay when the WPA2-802.1X standard is employed for authentication. (**a**) Average access delay versus n. (**b**) Average access delay versus β

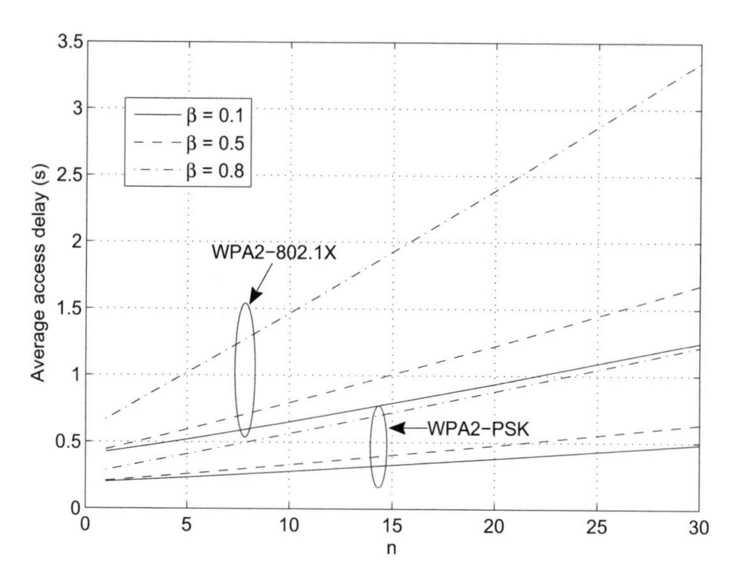

Fig. 2.5 Comparison of the access delay performance for the WPA2-802.1X and WPA2-PSK authentication mechanisms

Table 2.2 Testware of the experiment

Type	Hardware	Operating system	Software
WiFi AP	USB wireless adapter	Linux	Hostapd v2.6, Iperf, Wireshark
WiFi client	USB wireless adapter	Windows	Iperf
Authentication server	Virtual machine	Linux	FreeRadius v2.1
DHCP server	Virtual machine	Linux	ISC DHCP server

2.2.4 Experiment

To further study the accuracy of the analytical model in Sect. 2.2.2, we conduct experimental testing with COTS WiFi products and a cutting-edge channel emulator, to investigate the average access delay with different number of contending nodes, under realistic channel conditions in a vehicular environment, and by using the WPA2-802.1X and the WPA2-PSK authentication mechanisms. The experiment framework, test procedure, and test results are presented in the following.

As shown in Fig. 2.6, the experiment framework consists of a WiFi AP, multiple WiFi clients (representing the tagged vehicle and its neighbor vehicles), and a channel emulator. The testware of the experiment is summarized in Table 2.2, and each component of the experiment framework, as well as the testing procedure, is described as follows.

WiFi AP The WiFi AP functionalities are performed by using a universal serial bus (USB) wireless adapter, together with the hostapd software [37], which supports

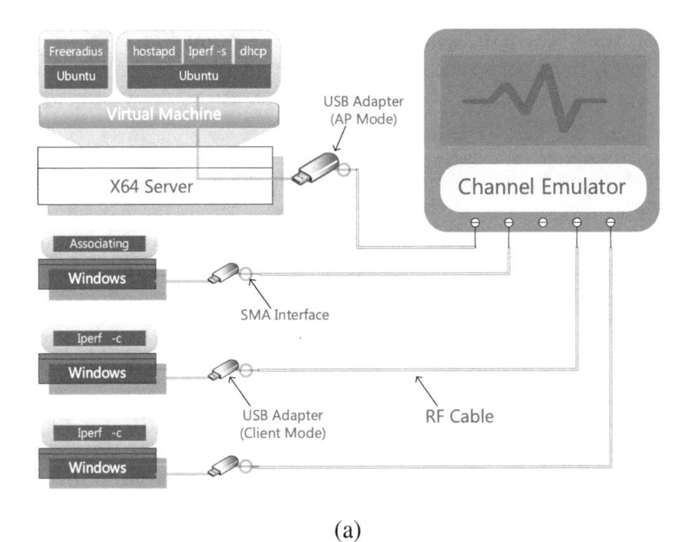

(a)

(b)

Fig. 2.6 Experiment framework. (**a**) Schematic diagram. (**b**) Testbed

WiFi AP operation on COTS wireless adapters. The AP operates over the 5.2 GHz WiFi channel, based on the high throughput (HT) PHY defined in the IEEE 802.11n amendment. While the IEEE 802.11n HT PHY supports multiple-input-multiple-output (MIMO) antenna configuration, only one antenna is used by the AP and each WiFi client, in order to reduce the number of physical connections from the wireless adapters to the channel emulator and simplify the emulation of the channel conditions among the wireless adapters.

WiFi Clients Similar to the WiFi AP, the WiFi client operation is achieved by using a USB wireless adapter. One of the WiFi clients represents the tagged vehicle, while the other clients represent the neighbor vehicles. The client representing the tagged vehicle is configured to connect to and then disconnect from the AP, continuously for the whole duration of each experiment. In each experiment, the other clients (neighbor vehicles) are set up to continuously generate and upload data frames to the AP, by using the iperf tool [38].

Authentication Server As shown in Fig. 2.2, the WPA2-802.1X mode requires communication between the AP and an AAA server in order to authenticate a WiFi client. Hence, in our experiments involving the WPA2-802.1X authentication mechanism, the Freeradius software is used to authenticate the WiFi client that represents the tagged vehicle, by applying the EAP and the tunneled transport layer security (TTLS). The Freeradius server runs on a virtual machine on the same computer that runs the hostapd server for the AP operation.

DHCP Server We use DHCP for the AP to automatically provide a valid IP address for a client to set up the Internet connection after authentication. The DHCP server runs on the same virtual machine as the hostapd server, as shown in Fig. 2.6a, and all the IP addresses are assigned on the same subnet as the AP interface.

Channel Emulator We use the Propsim F32 channel emulator, which emulates the effect of the wireless channel, such as noise, fading, delay, shadowing, and transceiver mobility, in order to conduct the experiments under realistic wireless channel conditions in a vehicular environment. The channel impulse response (CIR) is defined for the Propsim emulator in the form of a tapped delay line, where each tap represents a combination of line of sight (LoS) or non-LoS (nLoS) paths, through which the transmitted signal propagates to the destination. Each tap specifies the characteristics of the CIR component that is received through the propagation paths corresponding to the tap, such as the excess delay value, average received power, magnitude probability density function, and Doppler power spectral density (PSD). In our experiments, we use two channel models that are developed for vehicle-to-vehicle (V2V) and vehicle-to-roadside-unit (V2R) communications in the 5.9 GHz band in an expressway scenario [39].[6] The channel models used for communication between the AP and any vehicle, and between any two vehicles are illustrated in Fig. 2.7a and b, respectively, where the average power of the tap that involves the LoS path, referred to as the LoS tap (occurring at 0 excess delay), is normalized to unity. The magnitude of the LoS tap follows a Rician distribution, while the magnitude of each of the other taps follows a Rayleigh distribution. The Rician

[6] The two channel models under consideration are referred to in [39] as (1) RTV expressway and (2) V2V expressway same direction with wall.

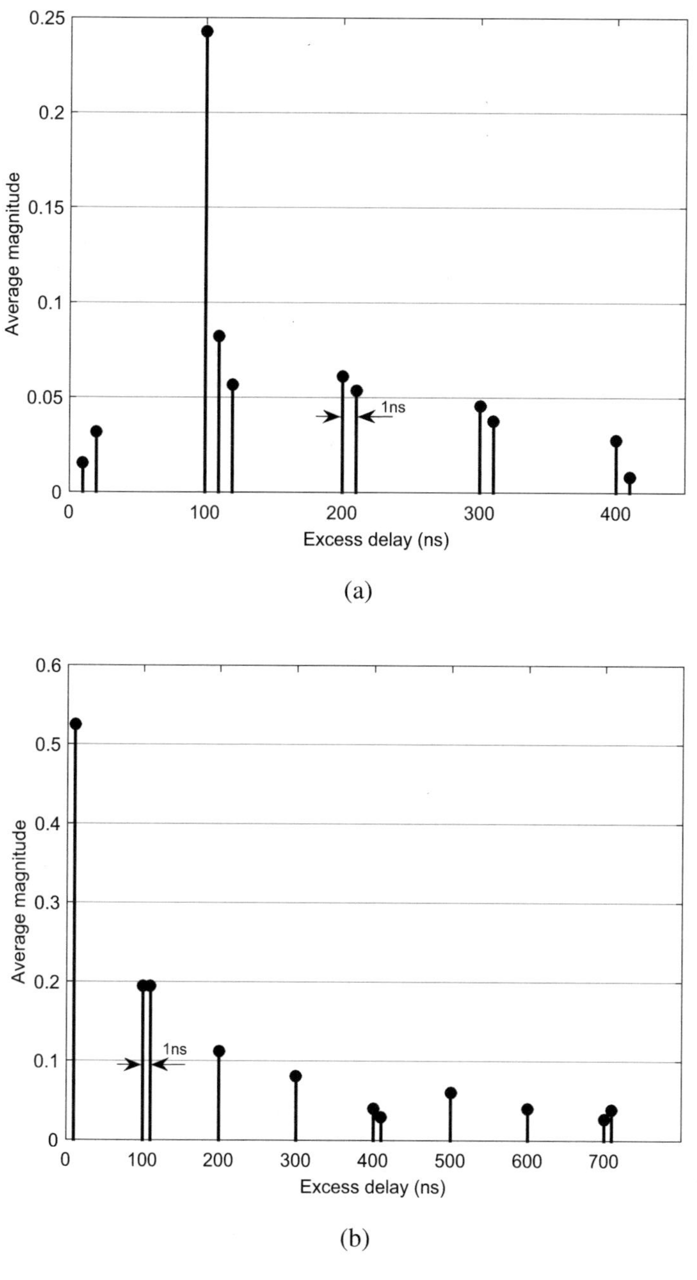

Fig. 2.7 The V2R and V2V channel models used in the experiments. The average magnitude of the LoS tap is normalized to unity (the LoS tap is not shown). (**a**) V2R channel model. (**b**) V2V channel model

Table 2.3 Test cases

Number of clients (n)	β	Authentication protocol
1, 2, and 3	0.3 and 0.4	WPA2-PSK and WPA2-802.1X

K-factor of the LoS tap and the Doppler PSD characteristics of all taps are specified in [39]. The signal received by the AP or any vehicle is corrupted by additive white Gaussian noise (AWGN), and the signal-to-noise ratio (SNR) is set to either 40 or 35 dB. These SNR values result in two different β values, as discussed next.

Test Procedure The test cases under consideration are summarized in Table 2.3. Given the channel model between the tagged vehicle and the AP, the values of β that correspond to the SNR values of 40 and 35 dB are found to be approximately 0.3 and 0.4, respectively, as indicated in Table 2.3. For each SNR value, the corresponding β value is estimated by conducting a separate test using the AP and a WiFi client that continuously sends data frames to the AP. The β value is estimated by calculating the ratio of the number of data frames received by the AP for the channel model and the SNR value under consideration to the number of received frames by the AP for an ideal channel (an option in the emulator) over the same time duration.

For each test case in Table 2.3, the tagged vehicle continuously connects to and then disconnects from the AP for a duration of 1 h. Each time the tagged vehicle completes a connection to the AP, the exchanged management frames are recorded using a Wireshark protocol analyzer, which captures all the frames transmitted or received by the AP.[7] The Wireshark generates a trace file that includes the contents of each captured management frame and a time stamp indicating the time instant the frame was received or transmitted by the AP. By analyzing the Wireshark trace file generated for a certain test case, the access delay can be obtained for each connection performed by the tagged vehicle to the AP, then the average access delay for the test case is calculated over all the connections achieved during the 1-h test duration. The parameter values employed for all test cases are summarized in Table 2.1, which are the same as those used to generate the numerical and simulation results in Sect. 2.2.3.

The performance of the access delay for the test cases under consideration is shown in Fig. 2.8. It can be seen from Fig. 2.8a and b that, for both the WPA-PSK and WPA2-802.1X authentication mechanisms, increasing the β value (from 0.3 to 0.4) does not result in a significant increase in the average access delay for the n values, which is consistent with the analytical result in Fig. 2.4b. Table 2.4 lists the difference between the average access delay values obtained from the analysis in Sect. 2.2.2 and the experiments. It is observed that, for all the n and β values under consideration, and for the two authentication mechanisms that we tested, there is a good match between the analytical and experimental results. The maximum difference between the analytical and experimental values of the average access

[7] The first four frames in Figs. 2.1 and 2.2 could not be captured in any experiment.

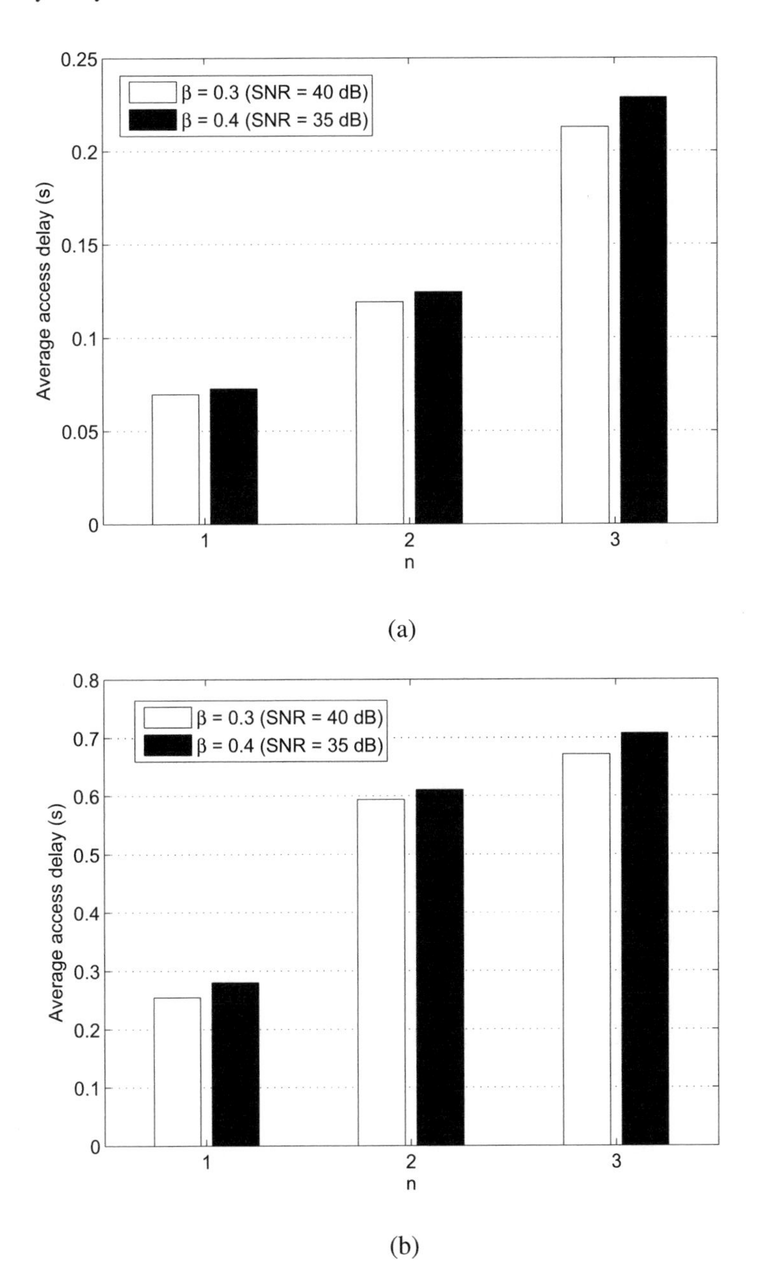

Fig. 2.8 Experimental results of the access delay when the WPA-PSK mode and the WPA2-802.1X standard are employed for authentication. (**a**) WPA2-PSK. (**b**) WPA2-802.1X

Table 2.4 Difference between the average access delay values obtained from the analysis in Sect. 2.2.2 and the experiments in Sect. 2.2.4

		n					
		1	2	3	1	2	3
β	0.3	0.135	0.094	0.011	0.175	0.137	0.186
	0.4	0.133	0.092	0.041	0.155	0.145	0.210
		WPA2-PSK			WPA2-802.1X		

delay is around 0.2 s. One reason of some mismatch between the analytical and experimental results is the additional (random) delay introduced by the channel emulator, which is not accounted for in the access delay calculation.

2.2.5 *Summary*

In this section, we have developed an analytical model to evaluate the access delay for a vehicle user to connect to the Internet through an on-road WiFi AP. The access delays of the WPA2-PSK and WPA2-802.1X authentication protocols are analyzed for different numbers of contending nodes and under various channel conditions (represented by frame error rates). It is shown that the access delay increases almost linearly with the number of contending nodes and the rate of increase is higher when the channel conditions result in a high frame error rate. Additionally, for a small number of contending nodes, increasing the frame error rate (e.g., up to 50%) does not result in a significant increase in the average access delay. It is also shown that the access delay of the WPA2-PSK authentication method is significantly less than that of the WPA2-802.1X, which highlights the importance of carefully selecting a suitable authentication method for WiFi access in a vehicular environment. The proposed analytical model and experiment framework can be applied to evaluate the access delay performance of newly developed authentication schemes.

2.3 Throughput Capacity Analysis of Drive-Thru Internet

In Sect. 2.2, the average access delay, defined as the duration to accomplish the access procedure is analyzed and measured, which can be regarded as the metric to evaluate the throughput performance of the drive-thru Internet, as higher delay will cause less Internet connection time, and thus reduce the data amount that transmitted between the vehicle and the AP. However, given the value of *average access delay*, it is still not enough to obtain the actual throughput performance since the WiFi link rate is varying with distance between the vehicle and the AP [14]. Following the traditional theory, the Probability Density Function (PDF) $f_a(t)$ is required to calculate the average throughput D_a, which is difficult to obtain:

$$D_a = \int_t^{\frac{L}{v}} \overline{\mathrm{R}}(vt, n) f_a(t)\, dt \tag{2.18}$$

where $\overline{\mathrm{R}}(x, n)$ is the average WiFi link rate when the distance between the vehicle and AP is x, and v, L are the vehicle velocity (assuming that vehicle velocity does not change in the coverage area) and the coverage range respectively and n is the number of co-associated WiFi clients. In [40] the management frame transmission during the access procedure is demonstrated as the state transition of a 2D Markov process as the vehicle drives through different zones of the coverage area. However, the factors to fulfill each management frame, such as number of co-associated WiFi clients, management frame drop rate due to channel error, etc., were not considered. Instead of calculating $f_a(t)$, we adopt a 3D Markov process to obtain the throughput performance by study the position transition upon the access procedure with the consideration of the transmission contention for each management frame, which is affected by the number of clients in the same cell, management packet drop rate due to channel loss, and the 802.11 back off process.

2.3.1 System Model

The system model is elaborated in the following parts, including the network model, zone transition model and Medium contention model.

As shown in Fig. 2.9, a stretch of road is covered by a WiFi AP located at the roadside. As soon as the vehicle drives into the coverage area where the Signal-to-Noise Ratio (SNR) grows to certain level, the access procedure is started, and before its accomplishment the vehicle cannot transmit any data frame to the AP. Two common kinds of access protocols are considered in the access procedure, i.e., the WPA2-PSK and Hotspot 2.0.[8] The total number of the management frames during the access procedure for the two protocols are different, and in following analysis it is denoted by N_A.

We consider two authentication protocols, i.e., WPA2-PSK and Hotspot 2.0. The comparison of the two access schemes are listed in Table 2.5.

Assuming there are already n WiFi clients (except the approaching vehicle, can be other vehicles or other types of users) associated to the same AP and keep on sending packets to it, i.e., under a saturated traffic situation. As the access procedure takes places in the edge of the network coverage area, the transmission of the management frames is likely to fail due to the unreliable channel condition, whose probability is denoted by β. According to 802.11 protocol, the link modulation rate

[8] The Hotspot 2.0 specifies more contents other than the access procedure, such as the QoS mapping, online sigh up, etc. [14].

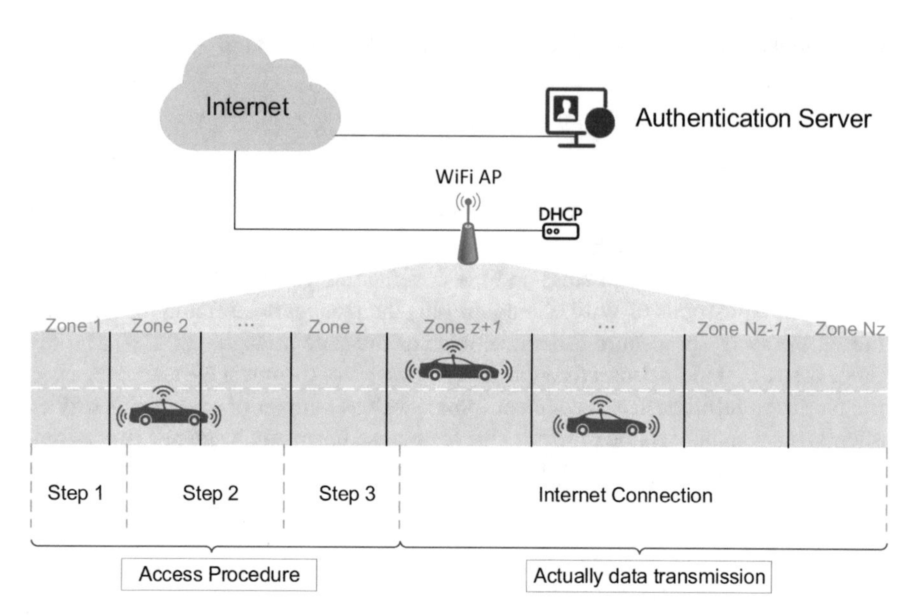

Fig. 2.9 System model

Table 2.5 Management frames of access protocols

Access scheme	Hotspot 2.0			WPA2-PSK		
Step	Protocol	N_A	Backhaul delay	Protocol	N_A	Backhaul delay
AP detection	ANQP handshake	2	No	Beaconing	2	No
Authentication	802.1X	27	Yes	PSK	8	No
IP address	DHCP	2	No	DHCP	2	No

depends on the SNR level, which is mainly determined by the distance between the AP and the vehicle [23, 41]. The link rate is assumed to be the same within a specific zone in the coverage area, which is defined in the following subsection.

We consider that the road stretch covered by the WiFi AP can be divided into N_z zones based on the modulation link rate between the vehicle and the AP. And as the vehicle drives through the coverage area, it traverses the zones consecutively. The duration the vehicle stays in an arbitrary zone z, denoted by t_z, equals to d_z/v, where d_z is the size of the zone. The link rate when the vehicle is in zone z is denoted by r_z. Similar to [23], the time a vehicle stay in each zone is approximated as an geometrical variable with mean of t_z. Within an relatively small duration δ, the probability that the vehicle in zone z will transit to the next zone $z + 1$ equals to δ/t_z, and two consecutive transitions are independent. If the vehicle velocity changes across different zones, the average sojourn duration inside each zone would also change, so does the corresponding zone transition probability. Without loss of generality, we assume that the velocity remains the same and t_z is determined solely by the value of v and sizes of zones.

To investigate the affect of co-transmission from contending WiFi clients, e.g., vehicles associated to the same AP, the media access procedure for each management frame is also investigated. We apply the 802.11 DCF without employing the RTS/CTS scheme, and the minimum contention windows size is denoted by w, while the number of the back off stages is denoted by m.[9]

2.3.2 3D Markov Chain Based Throughput Analysis

In this section, we adopt a 3D Markov chain to demonstrate the proceeding of the access procedure, which includes the dimensions of back off procedure, management frame index sequence and zone transition. We first build a 1D Markov chain for the back off stage transition when transmitting a certain management frame, and then expand it to a 2D Markov chain by considering the sequence of the management frames. And by sampling the beginning and ending moments of each status in this 2D Markov chain for the zone transition process, we finally form a 3D Markov chain model, which is used to calculate the relationship between the vehicle position and the accomplishment of access procedure, which is then used to obtain the drive-thru Internet throughput.

2.3.2.1 Dimension of Back Off Procedure

Consider the kth management frame during the access procedure, which should be generated before transmission. The frame generation involves the local protocol process, potential backhaul handshake, and network framing. The whole procedure is defined as the *core process*, after that the frame is sent based on the 802.11 DCF, which incurs the back off procedure with a random back off counter C_i at stage i, $i \in [0, m-1]$, and the initial stage i is set to 0:

(1) The back off counter C_i is uniformly selected from $[0, w * 2^i - 1]$, where w is the minimum window size.
(2) If the channel is sensed idle, then decrease the back off counter per time slot; if the channel is busy, then freeze the back off counter.
(3) If the C_i decreases to zero, then attempt to transmit the frame to air;
(4) If the transmission fails either due to collision or channel error, then increase the stage value i by one until to its maximum value, and go to step (1).[10]
(5) If the transmission is successful and the corresponding ACK is received, then jump to the frame generation (*core process*) of the next management frame.

[9] Different values of w and m may be employed to provide differential services, e.g., for delay-sensitive voice traffic [42], however we consider all WiFi clients use the same w and m values.

[10] There is no try limit to transmit a frame.

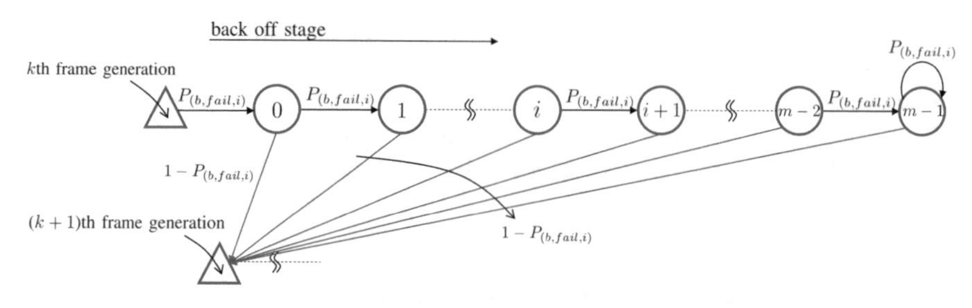

Fig. 2.10 Dimension of back off procedure

As the frame generation is independent with the frame transmission, and any transmission attempt is independent with the previous one, by sampling the end moment of each transmission attempt, the transition of the back off stage value i forms a discrete time Markov Chain [36], as demonstrated in Fig. 2.10, the triangle represents the status of *core process*, i.e., management frame generation, whose duration is denoted by $t_{c,k}$, while the circle represents the ith back off stage.

The one-step transit probability when transmitting kth management frame, denoted by $P_{(b,fail,k)}$ during the back off procedure can be calculated via the equations in Bianchi's method [36]:

$$\tau_k = \frac{2(1-2P_{(b,fail,k)})}{(1-2P_{(b,fail,k)})(w+1)+P_{(b,fail,k)}w(1-(2P_{(b,fail,k)})^{m-1})} \tag{2.19a}$$

$$\rho_k = 1-(1-\tau_k)^n \tag{2.19b}$$

$$P_{(b,fail,k)} = 1-(1-\rho_k)(1-\beta_k). \tag{2.19c}$$

where ρ_k is the probability of collision for a transmission attempt of the kth frame, β_k is the probability that the transmission of kth management frame fails due to channel error. Given current stage is i, the next stage will be $i+1$ if the transmission fails, otherwise it will directly jump to the *core process* of the $(k+1)$th management frame.

The average time during the status of the kth frame's *core process*, $\mathbb{E}(t_{c,k})$, depends on the access protocol and the backhaul network status, and can be measured from a real WiFi system [19]. The average time during each back off stage i can be calculated by

$$\mathbb{E}(T_i) = \mathbb{E}(T_{idle}) + \mathbb{E}(C_i) * \mathbb{E}(T_\sigma) + P_s * T_s + P_{us} * T_{fail} \tag{2.20}$$

where $\mathbb{E}(T_{idle})$ is average duration to sense the channel to be idle and then the back off procedure can be invoked, which can be approximated to the time for successfully transmitting a data frame in saturated condition:

$$\mathbb{E}(T_{idle}) = T_h + \frac{L_d}{r_{d,z}} + \text{SIFS} + \frac{L_{ack}}{r_{ack}} \tag{2.21}$$

T_h is the time to transmit the physical header section of the wireless frame, such as the PLCP field, PLME field, etc. L_d is the length of the data frame and is assumed to be identical for all WiFi clients. $r_{d,z}$ is the modulated link rate of the data frame when the vehicle is located at zone z. SIFS is the length of the Short Interframe Space as specified in 802.11 protocol. L_{ack} and r_{ack} are the length and the rate of the ACK frame respectively.

The average value of the back off counter at stage i in Eq. (2.20) can be calculated as below since the counter is uniformly distributed in $[0, w * 2^i - 1]$

$$\mathbb{E}(C_i) = \frac{1}{2}(w * 2^i - 1) \tag{2.22}$$

$\mathbb{E}(T_\sigma)$ represents the average time spent to decrease one back off counter. Denote the length of a idle time slot in 802.11 protocol by σ, and T_σ may include several σ and back off durations, which can be calculated by considering three conditions. First, in a given slot, it is possible that all the other vehicles are not transmitting in the given time slot, whose probability is denoted by p_0, and in this situation, the time to decrease one back off counter equals to σ. Secondly, it is also possible that there is only one vehicle is transmitting in the given time slot with the probability of p_1, and if the transmission is successful, then it requires $t_{1,s}$ to decrease on back off counter:

$$t_{1,s} = T_h + \frac{L_d}{r_{d,z}} + \text{DIFS} + \text{SIFS} + \frac{L_{ack}}{r_{ack}} \tag{2.23}$$

And if the transmission fails due to channel loss with the probability of β_d, there is no ACK frame and thus it requires $t_{(1,fail)}$ to decrease one back off counter:

$$t_{(1,fail)} = T_h + \frac{L_d}{r_{d,z}} + \text{DIFS} \tag{2.24}$$

Thirdly, if there are two or more other vehicles are transmitting, which means there will be transmission collision, whose probability is denoted by $p_{(2+)}$, then it also requires $t_{(1,fail)}$ to decrease one back off counter. And $\mathbb{E}(T_\sigma)$ can be calculated by

$$\begin{aligned}\mathbb{E}(T_\sigma) =& p_0\sigma + p_1 t_{1,s}(1-\beta_d) + t_{(1,fail)}[p_{(2+)} + p_1\beta_d] \\ =& (1-\rho_k)\sigma + p_1(1-\beta_d)(\text{SIFS} + \frac{L_{ack}}{r_{ack}}) \\ &+ (p_1 + p_{(2+)})t_{(1,fail)}\end{aligned} \tag{2.25}$$

where p_1 can be obtained by

$$p_1 = C_n^1 \tau_k (1 - \tau_k)^{n-1} \tag{2.26}$$

and $p_1 + p_{(2+)}$ equals to the probability that there are one or more other vehicles are transmitting, which is ρ_k.

The probability that the management frame is transmitted successfully, denoted by P_s in Eq. (2.20), equals to $1 - P_{(b,fail,k)}$, while P_{us} equals to $P_{(b,fail,k)}$. The air time spent to successfully transmit the management frame T_s can be obtained via

$$T_s = T_h + \frac{L_k}{r_k} + \text{SIFS} + \frac{L_{ack}}{r_{ack}} \tag{2.27}$$

where L_k is the length of the kth management frame. And the air time spent if the transmission fails T_{fail} can be obtained by

$$T_{fail} = T_h + \frac{L_k}{r_k} * \frac{\beta_k (1 - \rho_k)}{P_{b,fail,k}} + \frac{L_d}{r_{d,z}} * \frac{\rho_k}{P_{(b,fail,k)}} \tag{2.28}$$

where the second item in the above equation is an approximation of the air time to deliver the management frame given the condition that the transmission fails due to channel error, while the last item is the air time to deliver the data frame, as the transmission failure is caused by collision.[11]

2.3.2.2 Dimension of Management Frame Delivery Sequence

The frame generation and transmission of the $(k + 1)$th management frame only depends on the previous kth management frame, and thus the frame index, namely, k, forms a discrete Markov chain that takes the value from $[1, N_A]$, as shown in Fig. 2.11. The last 'accessed' status of '$N_A + 1$' indicates that all the management frames are exchanged, i.e., the access procedure has been accomplished. A similar model based on the consecutive delivery of the management frames can be found in [19], where the accuracy is verified both in numerical simulation and experiment.

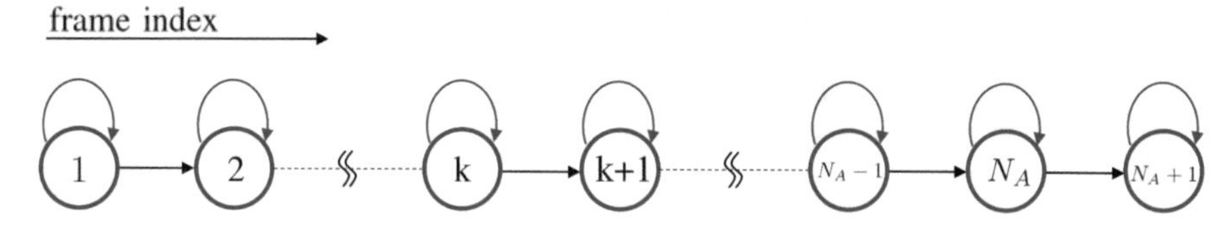

Fig. 2.11 Dimension of frame delivery sequence

[11] The length of the data frame is larger than all the management frame.

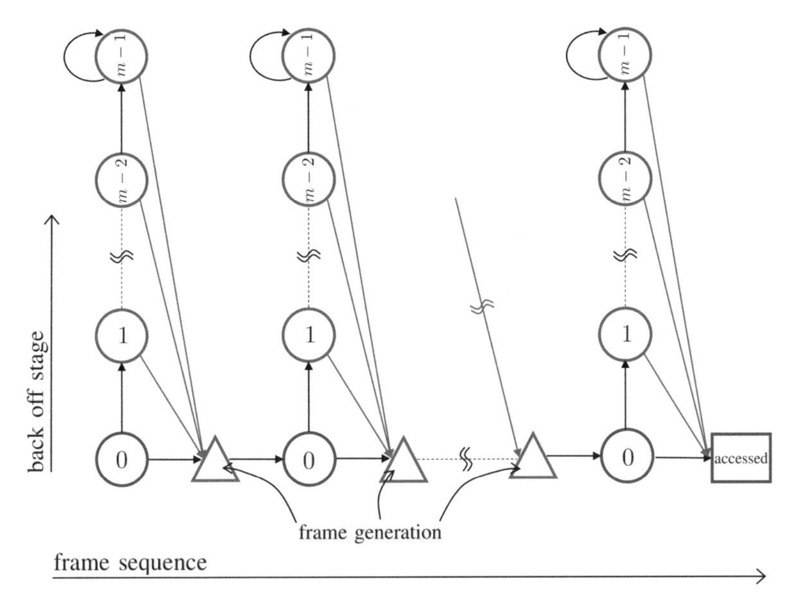

Fig. 2.12 2D Markov chain for back off procedure and frame delivery sequence

Since the transmission of each management frame is independent with each other, i.e., the back off stage altering of one frame is independent with another, thus, the Markov chain in Fig. 2.10 and the Markov chain in Fig. 2.11 together can form a 2D Markov chain of size $2N_A$ by m, which is shown in Fig. 2.12. In the dimension of the back off procedure, there are m status, while in the dimension of the frame sequence, there are $2N_A$ status, including the *core process* except the first frame (beaconing or ANQP query). And without loss of generality, the status of the vehicle can be denoted by a vector of (i, k) to indicate the transmission status of the management frame. The status of *core process* (in red triangles) indicates that the corresponding management frame is being generated.

The last status, demonstrated in the blue square at rightmost, indicates the status of the accomplishment of the access procedure. And the average duration for each status can be obtained as stated in Sect. 2.3.2.1.

2.3.2.3 Dimension of Zone Transition: Embedding 3D Markov Chain

With the assumption that the time of the vehicle in an arbitrary zone z follows the geometric distribution, and the probability to transit to next zone $z+1$ equals to δ/t_z for a short time δ, while the vehicle transverses through the coverage area, the zone index of the vehicle forms another Markov chain, as demonstrated in Fig. 2.13.

By adding the zone information to an arbitrary status in Fig. 2.12, we can describe the access procedure by three components: the back off stage i, the management

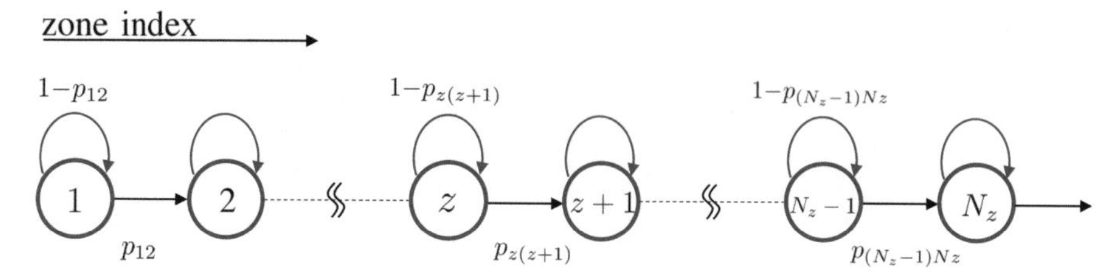

Fig. 2.13 Markov chain for zone transition

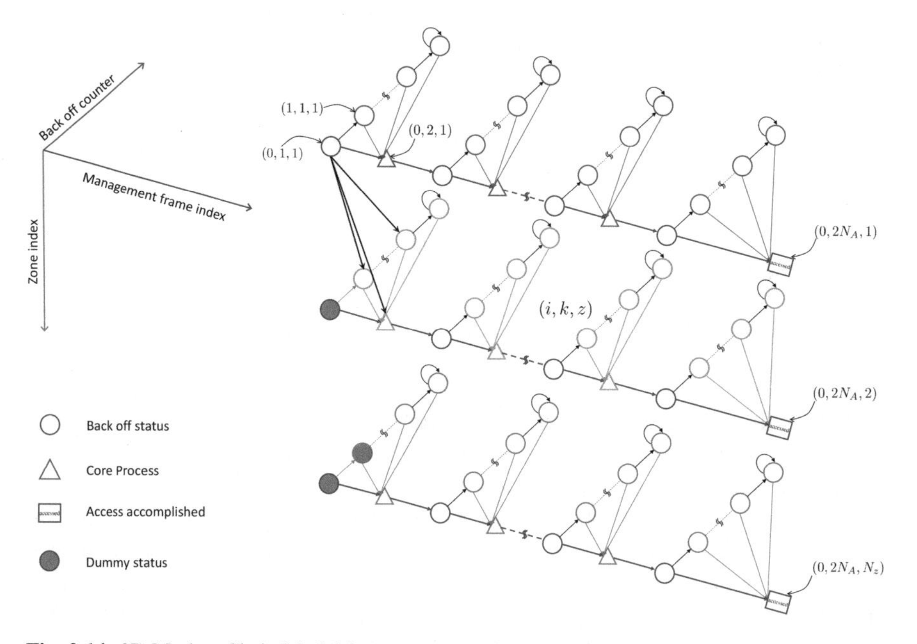

Fig. 2.14 3D Markov Chain Model for access procedure

frame index k and the zone index z, which are denoted by a vector of (i, k, z). As the zone transition depends solely on the mobility of the vehicle, and thus is independent with the transmission of every management frame, by sampling the beginning moment and the ending moment of each status, the vector (i, k, z) forms an embedded 3D Markov chain, which is demonstrated in Fig. 2.14. The 3D Markov chain includes following dimensions, which indicates the relationship with the accomplishment of the access procedure and the location (in which zone) of the vehicle.

(1) Back off stage dimension: the index i indicates the transmission of the currently management frame is in the back off process of stage i, i.e., the back off counter C_i is a uniformly distributed variable in $[0, w * 2^i - 1]$. The status begins at the ending moment of the previous transmission attempt and ends after the transmission attempt of this back off stage. In the *core process*, the frame is

under construction and has not triggered the back off process, so when k is even, the corresponding back off stage index equals to 0.

(2) Management frame index dimension: when the index k is odd, it represents the transmission of $(k+1)/2$th management frame; when the index k is even, it means the next management frame is being generated, and when k equals to $2N_A$, it means the whole access procedure is accomplished and the data traffic can be transmitted immediately.

(3) Zone index dimension: the index z indicates the location of the vehicle, i.e., the zone where the vehicle is located at the current moment. The smaller the zone index of the vehicle when finishing the access procedure, the earlier the vehicle can access to Internet, i.e., the higher the throughput can be achieved. The zone transition can possibly happen at the end of each status and the transition probability to the next zone depends on the average sojourn duration of the current status, i.e., given a status u while the vehicle is in zone z, then the probability that the next status will be in zone $z+1$ can be obtained by $P_u(z+1|z) = \mathbb{E}(T_u)/t_z$.

The one step transition probability of the 3D Markov chain can be obtained as follows. For a back off status (i, k, z) in Fig. 2.14, the next status after a transmission attempt can have four possibilities by considering the transmission result and the zone transition. If the transmission fails due to collision or channel error, while the vehicle stays in the same zone, then the next status will be $i+1, k, z$, or $m-1, k, z$ if the back off stage already reach the maximum value. The one step transition probability $\mathbb{P}(i+1, k, z|i, k, z)$ can be obtained by

$$\begin{aligned}\mathbb{P}(i+1, k, z|i, k, z) &= P_{(b,fail,k)}(1 - \frac{\mathbb{E}(T_{(i,k,z)})}{t_z}) \\ &= P_{(b,fail,k)}(1 - \frac{\mathbb{E}(T_i)}{t_z}), \\ i \in [0, m-2], k &\text{ is odd and} \in [1, 2N_A], z \in [1, N_z]\end{aligned} \tag{2.29}$$

and

$$\begin{aligned}\mathbb{P}(m-1, k, z|m-1, k, z) &= P_{(b,fail,k)}\left(1 - \frac{\mathbb{E}(T_{(m-1,k,z)})}{t_z}\right) \\ &= P_{(b,fail,k)}\left(1 - \frac{\mathbb{E}(T_{m-1})}{t_z}\right), \\ k \text{ is odd and} &\in [1, 2N_A], z \in [1, N_z]\end{aligned} \tag{2.30}$$

where $T_i, i \in [0, m-1]$ is the average duration when at the back off stage i, which can be obtained via Eq. (2.20). If $z \in [1, N_z - 1]$, i.e., not the last zone, and if transmission fails and the vehicle transits to next zone, then the next status will be $(i+1, k, z+1)$ (or $(m-1, k, z+1)$ if reaches maximum back off stage), whose probability can be obtained by

$$\mathbb{P}(i+1, k, z+1|i, k, z) = P_{(b,fail,k)}\frac{\mathbb{E}(T_i)}{t_z},$$
$$i \in [0, m-2], k \text{ is odd and } \in [1{,}2N_A], z \in [1, N_z - 1] \quad (2.31)$$

and

$$\begin{aligned}\mathbb{P}(m-1, k, z+1|m-1, k, z) &= P_{(b,fail,k)}\frac{\mathbb{E}(T_{(m-1,k,z)})}{t_z}\\ &= P_{(b,fail,k)}\frac{\mathbb{E}(T_{m-1})}{t_z},\end{aligned}$$
$$k \text{ is odd and } \in [1{,}2N_A], z \in [1, N_z - 1] \quad (2.32)$$

If the transmission is successful, then the next status will be the *core process* of the next management frame or the status of 'accessed' if all management frames are transmitted, i.e., $k = 2N_A - 1$. If the vehicle continues to stay in the current zone, then the next status will be $(0, k+1, z)$, and the one step transition probability can be obtained by

$$\mathbb{P}(0, k+1, z|i, k, z) = (1 - P_{(b,fail,k)})(1 - \frac{\mathbb{E}(t_{c,k})}{t_z}),$$
$$k \text{ is odd and } \in [1, 2N_A], z \in [1, N_z] \quad (2.33)$$

and if the vehicle moves to next zone, then

$$\mathbb{P}(0, k+1, z+1|i, k, z) = (1 - P_{(b,fail,k)})\frac{\mathbb{E}(t_{c,k})}{t_z},$$
$$k \text{ is odd and } \in [1, 2N_A], z \in [1, N_z - 1] \quad (2.34)$$

After the *core process*, the back off process is started as the management frame is ready to be transmitted. The next status will be the back off stage 0, and if the vehicle stays in current zone, the one step transition probability can be obtained by

$$\mathbb{P}(0, k+1, z|0, k, z) = (1 - P_{(b,fail,k)})(1 - \frac{\mathbb{E}(t_{c,k})}{t_z}),$$
$$k \text{ is even and } \in [1, 2N_A - 1], z \in [1, N_z] \quad (2.35)$$

and if the vehicle moves to next zone, then

$$\mathbb{P}(0, k+1, z+1|0, k, z) = (1 - P_{(b,fail,k)})(1 - \frac{\mathbb{E}(t_{c,k})}{t_z}),$$
$$k \text{ is even and } \in [1, 2N_A - 1], z \in [1, N_z - 1] \quad (2.36)$$

After the N_Ath management frame is exchanged, i.e., the access procedure has been accomplished, the next status will be 'accessed', as the squares shown in the right side of Fig. 2.14. The next status depends on the zone transition, we assume that for each 'accessed' status, the average duration is set to a relatively small value, which is denoted by $T_{(accessed)}$. And the one step transition probability can be obtained by

$$\mathbb{P}(0, 2N_A, z|0, 2N_A, z) = 1 - \frac{\mathbb{E}(T_{(accessed)})}{t_z},$$
$$z \in [1, N_z] \tag{2.37}$$

and

$$\mathbb{P}(0, 2N_A, z+1|0, 2N_A, z) = \frac{\mathbb{E}(T_{(accessed)})}{t_z},$$
$$z \in [1, N_z - 1] \tag{2.38}$$

When transmitting the first management frame, if the vehicle transits to the next zone after a transmission attempt, it will never enter back to the previous back off stage, i.e., the vehicle will not enter into the status of $(0, 1, 1)$, whose limiting probability equals to zero, which is defined as *dummy status*. Similarly, status of $(0, 1, 2)$, $(1, 1, 2)$, $(2, 1, 3)$, etc. are also *dummy status*. And hence, the number of overall status N_s is

$$N_s = \begin{cases} N_z(mN_A + N_A) - \frac{m}{2}(m-1), & m < z \\ N_z(mN_A + N_A) - \frac{z}{2}(z-1), & m \geq z \end{cases} \tag{2.39}$$

Assuming that as soon as the vehicle drives out of the WiFi network, it enters an identical WiFi coverage area and performs the access procedure again. Based on such renewal process, the expectation of the throughput of the drive-thru Internet can be obtained by letting the vehicle re-enters the same area infinitely and calculate the average value. Thus, for all status in the last zone in Fig. 2.14, i.e., $z = N_z$, the one step probability can be obtained by

$$\mathbb{P}(0, 1, 1|i, k, N_z) = \frac{\mathbb{E}(T_{(accessed)})}{t_z},$$
$$i \in [0, m-1], k \in [1, 2N_A], \tag{2.40}$$

Denote the limiting probability of an arbitrary status (i, k, z) by $\gamma_{i,k,z}$, i.e., the stable probability that the vehicle is transmitting the kth management frame at back off stage i and located in zone z. Given the one step transition probabilities Matrix, denoted by $\mathscr{M}$, formed by Eq. (2.29)–(2.40), together with the following uniform condition for the 3D Markov chain, the limiting probability vector γ can be obtained by

$$\begin{cases} \gamma \mathscr{M} = \gamma \\ \gamma e^T = 1 \end{cases} \tag{2.41}$$

The probability that the vehicle is located in a certain zone $\mathscr{Z}$ can be calculated by

$$P(\text{in Zone } \mathscr{Z}) = \sum_{z=\mathscr{Z}} \gamma_{i,k,z} \tag{2.42}$$

Given the vehicle is in zone $\mathscr{Z}$, the conditional probability that the access procedure is accomplished can be calculated by

$$P(\text{accessed}|\text{in Zone}\mathscr{Z}) = \frac{\gamma_{0,2N_A,\mathscr{Z}}}{P(\text{in Zone } \mathscr{Z})} \tag{2.43}$$

Denote the throughput a vehicle can achieved in zone z given the access procedure has been accomplished by $\mathscr{U}_z$, which can be obtained by

$$\mathscr{U}_z = \frac{r_z * t_z}{n+1} \tag{2.44}$$

where n is the number of the co-associated WiFi clients that share the bandwidth with the vehicle. And the overall throughput $\mathscr{U}_T$ can be calculated by

$$\mathscr{U}_T = \sum_{z=1}^{N_z} \mathscr{U}_z P(\text{accessed}|\text{in Zone } z) \tag{2.45}$$

2.3.3 *Simulation Results*

The simulation scenario includes a WiFi AP at roadside, a tagged vehicle that repeatedly drive over the coverage area, several WiFi clients that representing the co-associated contending vehicles, which keep on transmitting the data frames to AP. The WiFi AP adopts the 802.11n (HT) protocol [24] and the date rate is determined based on the free space path loss model for each zone [24, 43], whose parameters are listed in Table 2.6.

Table 2.6 The WiFi parameters in simulation

Parameters	Value
DCF time slot size σ	9 µs
SIFS size	16 µs
DIFS size	34 µs
PHY header duration (preamble+PLCP header)	20 µs
Data frame length	1574 bytes
ACK frame length	32 bytes
AP radio power	18 dbm
Frequency	5.2 GHz
Noise level	−95 dbm

According to the above specifications, the zone parameters can be calculated and are listed in Table 2.7. The link rate of the management frames are set to a constant value of 6 Mbps as observed from real access procedure, i.e., $r_k =$ 6 Mbps. The link rate of the ACK frames are the same with the corresponding both management frames and data frames. The length and the *core processing* delay for the management frames are measured from a real system employing the protocols of Hotspot 2.0 and WPA2-PSK similar with which in [19].

In one simulation run, the tagged vehicle will immediately perform the access procedure by exchanging the management set of a certain protocol, and upon its accomplishment, the vehicle start to continually transmit the data frame to the AP, and the overall data amount transmitted is record when it drives out of the coverage area, whose average value is considered to be throughput performance. And the above procedure is repeated for at least 200 runs to obtain the average throughput, which will be presented and discussed in the next sections.

Figure 2.15 shows the average aggregate throughput of the vehicle, which is obtained when the tagged vehicle can exclusively use all the link bandwidth after the access procedure. The maximum back off stage is set to 7 and the minimum window size is set to 16 according to the 802.11n (HT) protocol. The aggregate throughput shows the available overall throughput of the AP for the tagged vehicle's remaining journey after the access procedure. It can be observed that aggregated throughput is not degraded severely until the management frame drop rate reach to a significant value. The aggregated throughput decreases when there are more contending clients, which leads to more collisions and result in more transmission attempts, especially when the channel condition is worse. And comparing the two access schemes, the Hotspot 2.0 protocol adopts more management frames then WPA2-PSK, and also involves the backhaul delay, and thus lead to less aggregate throughput.

As in reality, the aggregate throughput will be shared by all the associated clients, as shown in Eq. (2.44). To demonstrate the impact the access procedure, the 'throughput loss' is defined to evaluate the difference of the throughput that can

Table 2.7 Zone parameters

Zone index	1	2	3	4	5	6	7	8	9	10	11	12	13	14	15	16	17
d_z (m)	26.8	23.9	8.4	12	6	3	5	2.4	8.2	2.4	5	3	6	12	8.4	23.9	26.8
r_z (Mbps)	6.5	13	19.5	26	39	52	58.5	65	78	65	58.5	52	39	26	19.5	13	6.5
$r_z * t_z$ (Mb) v=60 km/h	10.4	18.7	9.8	18.7	14	9.3	17.7	9.3	38.3	9.3	17.7	9.3	14	18.7	9.8	18.7	10.4

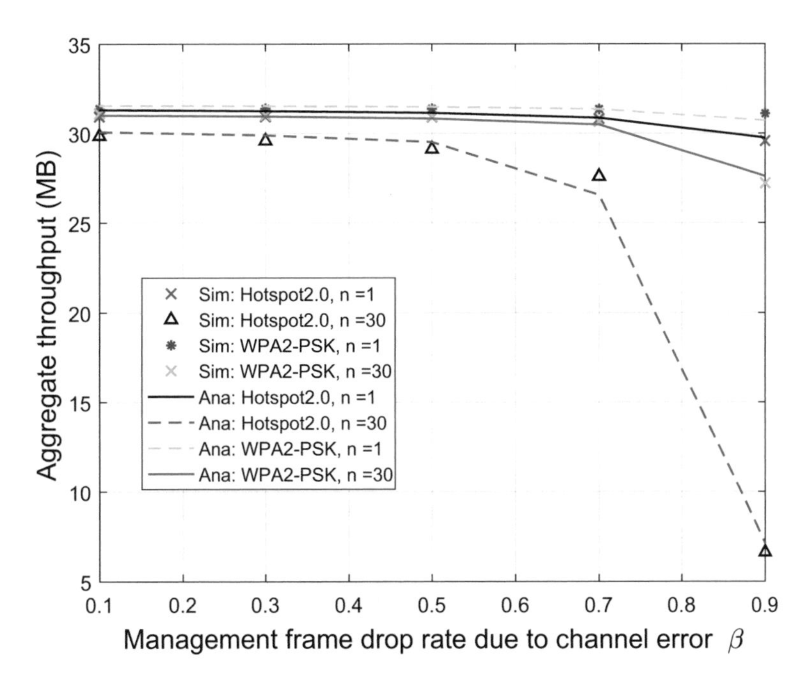

Fig. 2.15 Aggregate throughput vs management frame channel error rate

be achieved between the case of employing and not employing access procedure, which is denoted by η.

$$\eta = \frac{\mathcal{U}_{total} - \mathcal{U}_T}{\mathcal{U}_{total}} \tag{2.46}$$

where $\mathcal{U}_{total} = \sum_{z=1}^{N_z} \mathcal{U}_z$, which is the total throughput that the tagged vehicle can achieve without the access procedure. Figures 2.16 and 2.17 shows the throughput loss when employing the Hotspot 2.0 and the WPA2-PSK respectively. From the two figures, it is observed that when the number of the contending clients increase, the traffic loss is significantly exacerbated, especially when the frame drop rate is high, which is consistent with the results in Fig. 2.15. The traffic loss will be up to 80% when the Hotspot 2.0 is adopted, which means that the vehicle can only get around 20% of the expected throughput that no access procedure is employed. Similarly we can see a near 14% throughput loss when adopts WPA2-PSK, which requires to exchange much less management frames and no backhaul handshakes.

Figures 2.15, 2.16, and 2.17 show that the vehicle will lose significant throughput in the condition of large number of co-associated WiFi clients and high management frame drop rate due to channel error. Figure 2.18 shows that the throughput loss is related to the back off stage number, given a certain value (16) of minimum back

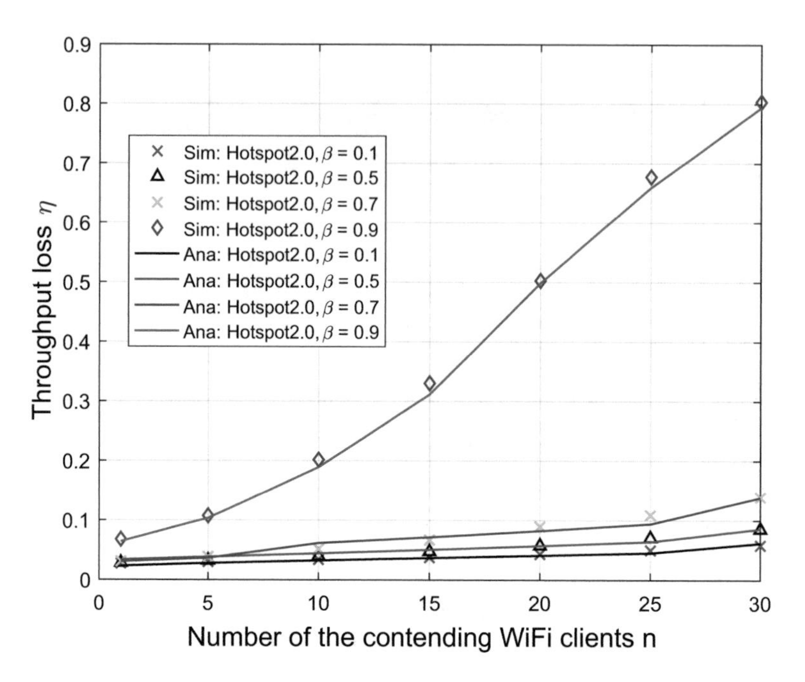

Fig. 2.16 Throughput loss of the tagged vehicle with Hotspot 2.0

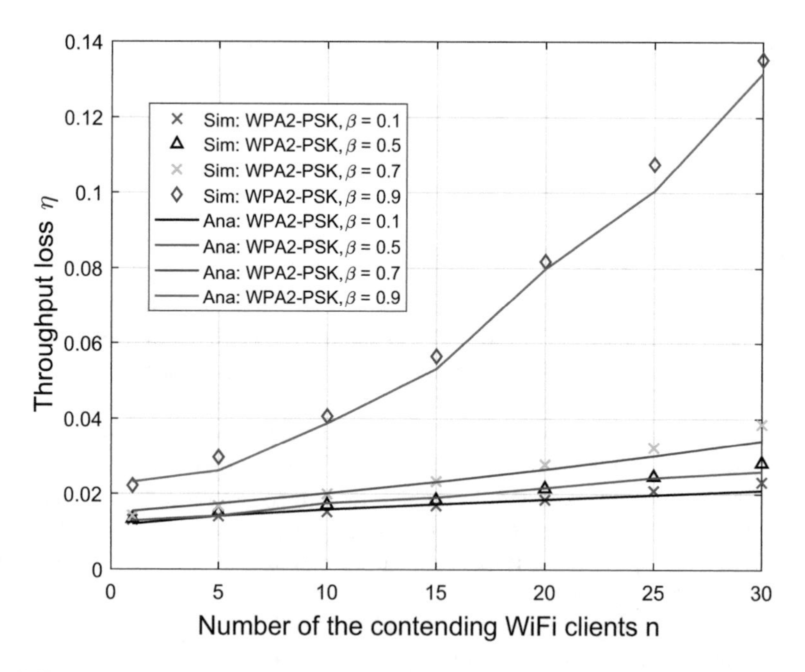

Fig. 2.17 Throughput loss of the tagged vehicle with WPA2-PSK

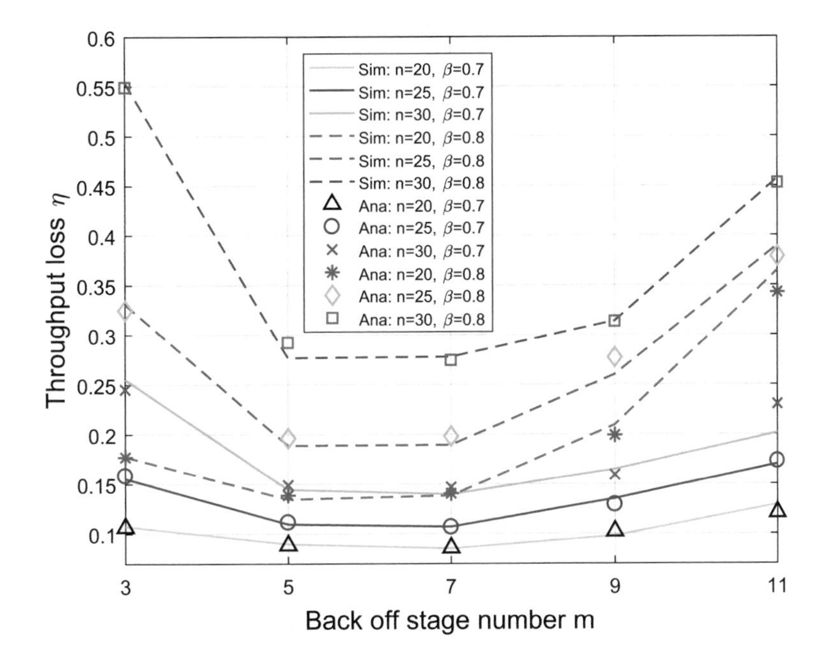

Fig. 2.18 Throughput loss vs. back off stage number m

off window size. When the back off stage number m become small, i.e., the back off stage i in Eq. (2.22) is limited to a small value, then the average back off counter C_i is reduced, and the vehicle will have less back off waiting time before a transmission attempt, and thus lead to high collision probability, which will obstruct the access procedure as the management frame transmission will be delayed. And when the back off stage number become larger, the average back off counter is increased, which means that for each transmission attempt, the average waiting time will be increased, and thus the overall back off waiting time will be increased, and the average time to deliver a management frame will also be increased, and thus the access procedure will consume more time.

A similar observation can be found in Fig. 2.19, which shows the traffic loss in the condition of different values of minimum back off window size w, given a certain back off stage number (7). When w is small, the back off time for a certain back off counter is limited, and thus the management frame will be attempted to transmit shortly, which will cause high collision probability in the similar way as the case of small m, and thus the duration to transmit a management frame will be increased, and the accomplish of the access procedure will need more time, which increase the traffic loss. While w is too large, the average waiting time for a given back off stage will be increased, and the average time used to transmit a management frame is thus increased, so more time will be used to finish the access procedure, which leads to high throughput loss as less Internet connection time will be available. Given a certain back off window parameter, it is possible to find the optimum back off stage

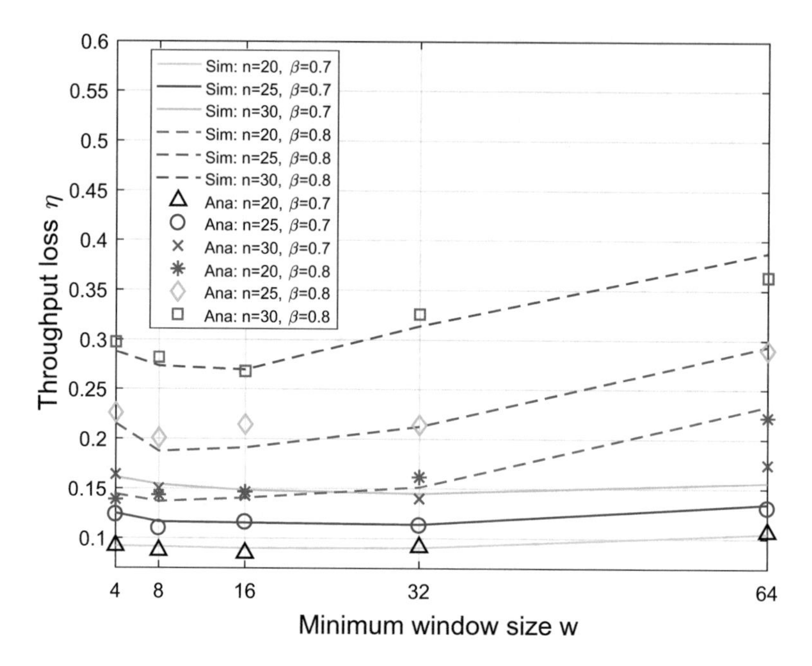

Fig. 2.19 Throughput loss vs. minimum window size

value that minimize the traffic loss from Fig. 2.18. And similarly give a certain back off stage parameter, the optimum back off window value can be found via Fig. 2.19. And in certain conditions, it is possible to find an optimal pair of the two parameters to minimize the traffic loss introduced by the access procedure by exhaust all values in these two figures.

2.3.4 *Summary*

In this section, we have investigated the access procedure for a vehicle to access to the roadside WiFi network for Internet services. We have proposed a 3D Markov chain model to calculate the throughput performance of drive-thru Internet by considering the transition of the coverage zones, management frame index and the back off process. We have conducted extensive simulations to validate the accuracy of our analytical model, which can be applied in the development of future vehicular networks, such as group authentication, mobile MAC protocol, software defined vehicular networks and connected autonomous vehicles.

References

1. N. Lu, N. Cheng, N. Zhang, X. Shen, J.W. Mark, Connected vehicles: Solutions and challenges. IEEE Internet Things J. **1**(4), 289–299 (2014)
2. W. Xu, H. Zhou, N. Cheng, F. Lyu, W. Shi, J. Chen, X. Shen, Internet of vehicles in big data era. IEEE/CAA J. Autom. Sinica **5**(1), 19–35 (2018)
3. H.A. Omar, W. Zhuang, L. Li, VeMAC: A TDMA-based MAC protocol for reliable broadcast in VANETs. IEEE Trans. Mobile Comput. **12**(9), 1724–1736 (2013)
4. F. Lyu, H. Zhu, H. Zhou, W. Xu, N. Zhang, M. Li, X. Shen, SS-MAC: A novel time slot-sharing MAC for safety messages broadcasting in VANETs. IEEE Trans. Veh. Technol. **67**(4), 3586–3597 (2018)
5. A. Meola, How the internet of things will transform private and public transportation. http://uk.businessinsider.com/internet-of-things-connected-transportation-2016-10. Accessed 2 Apr 2018
6. Y. Wu, L.P. Qian, H. Mao, X. Yang, H. Zhou, X. Shen, Optimal power allocation and scheduling for non-orthogonal multiple access relay-assisted networks. IEEE Trans. Mobile Comput. **17**(11), 2591 (2018)
7. K. Abboud, H.A. Omar, W. Zhuang, Interworking of DSRC and cellular network technologies for V2X communications: A survey. IEEE Trans. Veh. Technol. **65**(12), 9457–9470 (2016)
8. H. Zhou, N. Cheng, N. Lu, L. Gui, D. Zhang, Q. Yu, F. Bai, X. Shen, Whitefi infostation: Engineering vehicular media streaming with geolocation database. IEEE J. Sel. Areas Commun. **34**(8), 2260–2274 (2016)
9. Y. Wu, L.P. Qian, J. Zheng, H. Zhou, X. Shen, Green-oriented traffic offloading through dual connectivity in future heterogeneous small cell networks. IEEE Commun. Mag. **56**(5), 140–147 (2018)
10. A.K. Ligo, J.M. Peha, P. Ferreira, J. Barros, Throughput and economics of DSRC-based internet of vehicles. IEEE Access **6**, 7276–7290 (2018)
11. P. Luo, Z. Ghassemlooy, H. Le Minh, E. Bentley, A. Burton, X. Tang, Performance analysis of a car-to-car visible light communication system. Appl. Opt. **54**(7), 1696–1706 (2015)
12. Cisco, The zettabyte era: Trends and analysis. https://www.cisco.com/c/en/us/solutions/collateral/service-provider/visual-networking-index-vni/vni-hyperconnectivity-wp.pdf. Accessed 2 Apr 2019
13. E.H. Ong, J. Kneckt, O. Alanen, Z. Chang, T. Huovinen, T. Nihtilä, IEEE 802.11 ac: Enhancements for very high throughput wlans, in *2011 IEEE 22nd International Symposium on Personal Indoor and Mobile Radio Communications (PIMRC)* (2011), pp. 849–853
14. W. Xu, H. Zhou, Y. Bi, N. Cheng, X. Shen, L. Thanayankizil, F. Bai, Exploiting hotspot-2.0 for traffic offloading in mobile networks. IEEE Netw. **32**(5), 131–137 (2018)
15. J. Ott, D. Kutscher, Drive-thru Internet: IEEE 802.11 b for automobile users, in *IEEE INFOCOM 2004* (2004), p. 1
16. R. Mahajan, J. Zahorjan, B. Zill, Understanding WiFi-based connectivity from moving vehicles, in *Proc. ACM SigCom* (2007), pp. 321–326
17. N. Cheng, N. Lu, N. Zhang, X. Shen, J.W. Mark, Opportunistic WiFi offloading in vehicular environment: A queueing analysis, in *IEEE Global Communications Conference (GLOBECOM)* (2014), pp. 211–216
18. H. Zhou, B. Liu, T.H. Luan, F. Hou, L. Gui, Y. Li, Q. Yu, X. Shen, Chaincluster: Engineering a cooperative content distribution framework for highway vehicular communications. IEEE Trans. Intell. Transp. Syst. **15**(6), 2644–2657 (2014)
19. W. Xu, H.A. Omar, W. Zhuang, X. Shen, Delay analysis of in-vehicle internet access via on-road WiFi access points. IEEE Access **5**, 2736–2746 (2017)
20. Hotspot 2.0 technical task group, *Wi-Fi Alliance Technical Committee*
21. V. Bychkovsky, B. Hull, A. Miu, H. Balakrishnan, S. Madden, A measurement study of vehicular internet access using in situ Wi-Fi networks, in *Proc. ACM Mobile Computing and Networking* (2006), pp. 50–61

22. N. Cheng, N. Lu, N. Zhang, X. Zhang, X. Shen, J.W. Mark, Opportunistic WiFi offloading in vehicular environment: A game-theory approach. IEEE Trans. Intell. Transp. Syst. **17**(7), 1944–1955 (2016)
23. T.H. Luan, X. Ling, X. Shen, MAC in motion: Impact of mobility on the MAC of drive-thru internet. IEEE Trans. Mobile Comput. **11**(2), 305–319 (2012)
24. IEEE standard for information technology–telecommunications and information exchange between systems local and metropolitan area networks–specific requirements part 11: Wireless LAN medium access control (MAC) and physical layer (PHY) specifications, in *IEEE Std 802.11-2012 (Revision of IEEE Std 802.11-2007)* (2012), pp. 1–2793
25. S. Shin, A.G. Forte, A.S. Rawat, H. Schulzrinne, Reducing MAC layer handoff latency in IEEE 802.11 wireless LANs, in *ACM Proceedings of the Second International Workshop on Mobility Management & Wireless Access Protocols* (2004), pp. 19–26
26. J. Eriksson, H. Balakrishnan, S. Madden, Cabernet: vehicular content delivery using WiFi, in *Proc. of the 14th ACM International Conference on Mobile Computing and Networking* (2008), pp. 199–210
27. A. Tufail, M. Fraser, A. Hammad, K.K. Hyung, S.-W. Yoo, An empirical study to analyze the feasibility of WiFi for VANETs, in *12th International Conference on Computer Supported Cooperative Work in Design* (2008), pp. 553–558
28. C.-M. Chou, C.-Y. Li, W.-M. Chien, K.-c. Lan, A feasibility study on vehicle-to-infrastructure communication: WiFi vs. WiMAX, in *10th International Conference on Mobile Data Management: Systems, Services and Middleware, MDM'09* (2009), pp. 397–398
29. J. Wu, P. Fan, A survey on high mobility wireless communications: Challenges, opportunities and solutions. IEEE Access **4**, 450–476 (2016)
30. WiFi Alliance, Wi-Fi protected access: Strong, standards-based, interoperable security for today's Wi-Fi networks, in *White Paper, University of Cape Town* (2003), pp. 492–495
31. Hotspot 2.0 specification and passpoint project. [Online]. Available http://www.wi-fi.org/discover-wi-fi/wi-fi-certified-passpoint
32. J.P. Craiger et al., 802.11, 802.1 x, and wireless security, *SANS Institute InfoSec Reading Room* (2002)
33. J.-C. Chen, Y.-P. Wang, Extensible authentication protocol (EAP) and IEEE 802.1 x: tutorial and empirical experience. IEEE Commun. Mag. **43**(12), supl–26 (2005)
34. K. Ramezani, E. Sithirasenan, K. Su, Formal security analysis of EAP-ERP using casper. IEEE Access **4**, 383–396 (2016)
35. E.H. Ong, J. Kneckt, O. Alanen, Z. Chang, T. Huovinen, T. Nihtilä, IEEE 802.11 ac: Enhancements for very high throughput WLANs, in *2011 IEEE 22nd International Symposium on Personal, Indoor and Mobile Radio Communications* (IEEE, Piscataway, 2011), pp. 849–853
36. G. Bianchi, Performance analysis of the IEEE 802.11 distributed coordination function. IEEE J. Sel. Areas Commun. **18**(3), 535–547 (2000)
37. hostapd: IEEE 802.11 ap, ieee 802.1x/wpa/wpa2/eap/radius authenticator. http://w1.fi/hostapd/
38. iperf - the ultimate speed test tool for tcp, udp and sctp. https://iperf.fr/
39. M.A. Ingram, Six time-and frequency-selective empirical channel models for vehicular wireless LANs. IEEE Veh. Technol. Mag. **2**(4), 4–11 (2007)
40. W. Xu, H. Zhou, W. Shi, F. Lyu, X. Shen, Throughput analysis of in-vehicle internet access via on-road WiFi access points, in *Prof. IEEE VTC-Fall* (2017), pp. 1–5
41. L. Cheng, B.E. Henty, D.D. Stancil, F. Bai, P. Mudalige, Mobile vehicle-to-vehicle narrow-band channel measurement and characterization of the 5.9 GHz dedicated short range communication (DSRC) frequency band. IEEE J. Sel. Areas Commun. **25**(8), 1501 (2007)
42. P. Wang, H. Jiang, W. Zhuang, Capacity improvement and analysis for voice/data traffic over WLANs. IEEE Trans. Wireless Commun. **6**(4), 1530 (2007)
43. A. Goldsmith, *Wireless Communications* (Cambridge University Press, Cambridge, 2005)

Chapter 3
V2X Interworking via Vehicular Internet Access

In this chapter, we considered to utilize the V2V and V2R communication simultaneously to improve the Internet access performance. To strike a balance between the throughput and the time delay for a data tasks, we propose queueing based theory to analyze the tradeoff between the data volume that can be transferred and the time needed. We have analyzed the unique characteristics of the V2V and V2R communication, and apply the M/G/1/K queue to analyze their property in fulfilling the data tasks, which is used to conduct the interworking of V2V and V2R communication for the vehicular offloading, which is shown to significantly improve the Internet access performance.

3.1 Background and Motivation

Modern vehicles are equipped with sophisticated radio interfaces to connect to infrastructures (Vehicle-to-Infrastructure, V2I) and other vehicles (V2V) [1] to support the big data transmission generated by various automotive applications and services to enhance the safety level and efficiency of modern transportation [1]. it is predicted that the global vehicular traffic will reach 300 Zettabytes by 2020 [2, 3], which will be far beyond the capacity of cellular network, that will not only cause network congestion, but can also cause huge communication costs to vehicle users. Offloading the big data traffic to economic unlicensed networks, e.g., WiFi networks, are considered as a solution to reduce the communication cost while improve the throughput. In terms of utilizing the roadside WiFi network resource, Ott et al. conducted the well-known drive-thru Internet experiment, which indicated considerable UDP & TCP throughput can be achieved for drive-by vehicles [4]. Mahajan et al. measured the wireless connectivity between vehicles on road and APs along roadside, as well as the throughput performance of the WiFi connection [5]. Dimatteo et al. showed that the data delivery performance can be significantly

W. Xu et al., *Internet Access in Vehicular Networks*,
https://doi.org/10.1007/978-3-030-88991-3_3

improved even in sparse network deployment via WiFi offloading for vehicles [6]. Some research works from literature also demonstrate that the direct connection between mobile users can be utilized to further enhance WiFi offloading. Han et al. proposed a 'tagged user' selection scheme, and let tagged users to download the popular contents and then opportunistically share them to encountered peers through direct WiFi or Bluetooth connection [7]. Wu et al. utilized the cooperation between mobile users to allow mobile users to download the target contents and share with each other via the end-to-end connection [8]. Instead of offloading the common contents shared by some users, Cheng et al. analyzed the performance of the vehicular traffic offloading via intermittent roadside APs and studied the relationship between the offloading efficiency and the average data service delay given certain buffer length [9]. In vehicular conditions, it is also indicated that the V2V communication could help sharing content from the 'seed user' to others [10]. Since a vehicle is allowed to associate to only one AP at a time, when multiple APs are deployed along the roadside, there is a great potential to share the AP resources via the V2V connection based on the cooperation of the vehicles that associated to different APs.

In this chapter, we consider utilizing both the V2R and V2V communication to do the traffic offloading for vehicle users. We first setup a queueing model to evaluate the V2V data pipe's communication capacity, i.e., the relationship between the transmission delay and the offloading efficiency. Then we evaluate the way to utilize the interworking of V2V and V2R, i.e., to simultaneously use the two categories of communications to improve the offloading efficiency. We first setup an M/G/1/K queue model to describe the latency and service rate of the vehicular tasks, and investigate the relationship between various conditions and the offloading performance.

3.2 Queueing Model for Opportunistic V2V Assistance

3.2.1 System Model

We consider a set of vehicles that are traveling in urban area independently, i.e., without making a fleet or form a cluster. The nearby vehicles can setup V2V connection and transmit data content to each other and only the one-hop communication is considered. If the two vehicles drives out of the V2V range, the V2V connection will be interrupted. The network model and the queueing model are described as follows.

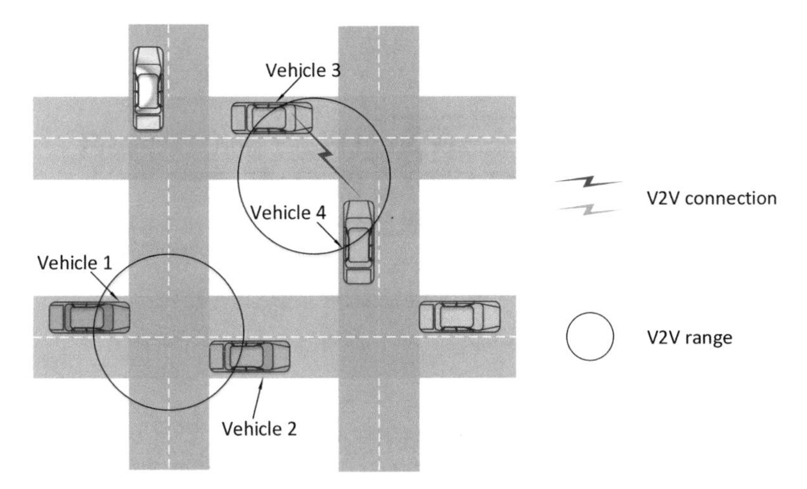

Fig. 3.1 V2V communication system model

3.2.1.1 Network Model

We assume that the radio interface installed on each vehicle can detect peers who are within the V2V communication rage and can transmit data content to one of them at a time, as shown in Fig. 3.1. The V2V connection would be interrupted in situation such as out of V2V range, blocking Line-Of-Sight (LOS) [11], etc. The life cycle of a V2V connection follows exponential distribution, as shown in Sect. 4.3.3, i.e., the V2V interruption is Poisson arrival.

The requirement to transmit a data content via the V2V connection to another peer is defined as one data task, e.g., sharing a message to peer, letting neighbor to relay content to remote destination, downloading data via peer's Internet access, etc. The time to fulfill a data task is a random variable and depends on the size of the content, V2V link rate and the interruption frequency and duration, etc. We assume that the efficiency of the V2V data pipe is quasi-stationary, i.e., the transmission time required for a certain data task stays the same during a certain time for every V2V connection, which is generally distributed and is denoted by t.

3.2.2 *Queueing Model*

The V2V data pipe is modeled as a server whose service time for an arbitrary data task is t with PDF of $f(t)$. To serve the Poisson arrival data tasks, we apply finite-size buffer to cache the arrived data tasks and queue for the V2V data pipe server to process with First-in-First-out discipline.

As shown in Fig. 3.2, the data transmission will stop when the V2V connection is interrupted, i.e., the server goes to vacation until the vehicle finds another peer

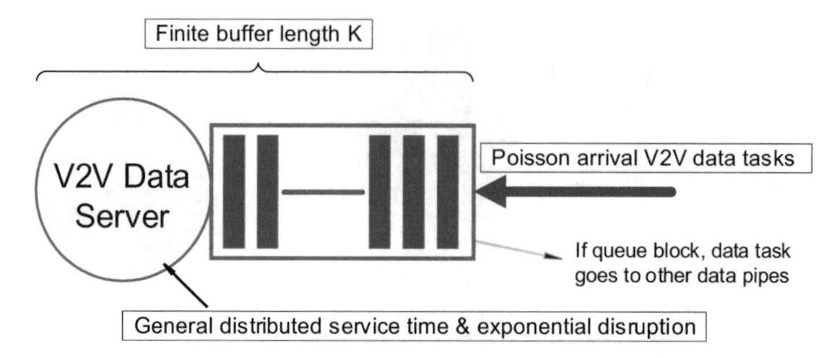

Fig. 3.2 V2V communication system model

to rebuild a new V2V connection or recover the previous V2V communication. The arrival rate of the interruption is denoted by i, and the time duration from the beginning moment that starting to transmit the data to the next interruption moment is denoted by I. Thus I is a exponential variable with mean of $1/i$, which is also referred to as the 'on-period'. In terms of the 'off-period' that the duration from the interrupted moment to the next V2V connection establishment, which is denoted by D and is generally distributed with PDF of $f_D(D)$, which is approximated from vehicle mobility trace, as shown in Sect. 4.3.3.

When a data task arrives and there is still task buffer space, it can enter into the buffer queue and wait for the V2V transmission. While if the queue is blocked, i.e., there are already K data tasks that occupy all space of the queue, the arriving data task cannot enter into the queue and thus would not be served, e.g., dropped or transmit the content via cellular connection, etc. The queue blocking probability, P_B, shows the effectiveness of the V2V data pipe, as it indicates the proportion of the data tasks that would not be fulfilled via V2V server among all arrived tasks. By adjusting the buffer length, we can do trade-off between how much V2V data tasks can be fulfilled and the average service time for them. A large K value would hold more data tasks while result in long queue size and thus lead to long waiting time, while a small K would drop more tasks however can reduce the waiting time in the queue.

3.2.3 Queueing Analysis About V2V Communication

The data content of a task may include several 'on-periods' and 'off-periods', since the transmission would be interrupted and sometimes cannot be fulfilled before the service interruption, whose total time is defined as the *effective service time* To obtain the value of P_B we can solve the queue status by calculating the distribute of the *effective service time*, which can be obtained via the Laplace Transform (L.T.) method [12]. For a data task which would consume time t 'on-period' time, then the *effective service time* can be denoted by $F(t)$, which has two different formats for different task resuming situations, namely, the SOR and BPR types. We first give the expression of the *effective service time* for both two types, and calculate its distribution via the L.T. method. Then based on its distribution, we solve an M/G/1/K queue to calculate the effectiveness of the opportunistic V2V communication and the average time used to serve the data tasks.

3.2.3.1 L.T. Method for Effective Service Time

We first give the relationship between the 'on-period' time t for a data tasks and the actual *effective service time* for two resuming types, and then calculate the L.T. of the *effective service time*.

3.2.3.2 SOR Type

For the SOR type, when the task is fulfilled before the arrival of next interruption, then the *effective service time* equals to the 'on-period' time required by the data task; otherwise, when the transmission restarts after a new V2V connection is setup since last interruption, the data task requires to re-transmit from the very beginning bit of the content, and thus we can have a recursive format for $F(t)$:

$$F(t) = \begin{cases} t, & if \quad t < I \\ I + D + F(t), & if \quad t \geq I. \end{cases} \tag{3.1}$$

The L.T. of $F(t)$ can thus be calculated by considering the above two cases:

$$L_F(s|t) = \mathbb{E}(e^{-sF}|t) \tag{3.2a}$$

$$= P(t < I)\mathbb{E}(e^{-sF}| \text{ given } t, \text{ and } t < I) \tag{3.2b}$$

$$+ P(t \geq I)\mathbb{E}(e^{-sF}| \text{ given } t, \text{ and } t \geq I) \tag{3.2c}$$

where s is a complex variable. The item in Eq. (3.2b) is about the case that the overall transmission can be finished before V2V interruption, which equals to

$$e^{-st}\int_t^{+\infty} f_I(x)dx = e^{-(i+s)t}. \tag{3.3}$$

While the item in Eq. (3.2c) equals to

$$\int_0^t f_I(I)\int_0^{+\infty} f_D(D)\int_0^{+\infty} f_F(F)e^{-s(I+D+F)}dFdDdI \tag{3.4a}$$

$$= \int_0^t ie^{-(i+s)I} dI \int_0^{+\infty} f_D(D)e^{-sD} dD \int_0^{+\infty} f_F(F)e^{-sF} dF \tag{3.4b}$$

$$= \frac{i}{i+s}(1 - e^{-(i+s)t}) L_D(s) L_F(s) \tag{3.4c}$$

where $L_D(s)$ is the L.T. of the random duration of 'off-period'. And then we have

$$L_F(s|t) = e^{-(i+s)t} + \frac{i}{i+s}(1 - e^{-(i+s)t}) L_D(s) L_F(s) \tag{3.5}$$

Since we have

$$\int_0^{+\infty} e^{-(i+s)} f_t(t) dt = L_t(i+s), \tag{3.6}$$

where $L_t(s)$ is the L.T. of the 'on-period' duration of a data task. And integrate all t for $L_F(s|t)$ in Eq. (3.5), we have

$$L_F(s) = \int_0^{+\infty} L_F(s|t) f(t) dt \tag{3.7a}$$

$$= L_t(i+s) + \frac{i}{i+s}(1 - L_t(i+s)) L_D(s) L_F(s) \tag{3.7b}$$

Then the L.T. of *effective service time* can be obtained:

$$L_F(s) = \frac{(i+s) L_t(i+s)}{i + s + i L_D(s)(1 - L_t(i+s))} \tag{3.8}$$

BPR Type For the BPR type, when the new V2V connection is established after an interruption, only the remaining part of the data content is required to transmit, and thus for data tasks that are interrupted, $F(t)$ changes to:

$$F(t) = \begin{cases} t, & if \quad t < I \\ I + D + F(t - I), & if \quad t \geq I. \end{cases} \tag{3.9}$$

The L.T. of $F(t)$ can be obtain similarly following Eq. (3.2), note that the item in Eq. (3.2b) does not change, while the item in Eq. (3.2c) changes to:

$$\int_0^t f_I(I) \int_0^{+\infty} f_D(D) \int_0^{+\infty} f_F(F|t - I) e^{-s(I+D+F)} dF dD dI \tag{3.10a}$$

$$= \int_0^t ie^{-(i+s)I} \int_0^{+\infty} f_D(D) e^{-sD} dD \int_0^{+\infty} f_F(F|t - I) e^{-sF} dF dI \tag{3.10b}$$

$$= iL_D(s) \int_0^t e^{-(i+s)I} L_F(s|t-I)dI \tag{3.10c}$$

$$= iL_D(s) \int_0^t e^{-(i+s)(t-r)} L_F(s|r)dr \tag{3.10d}$$

$$= ie^{-(i+s)t} L_D(s) \int_0^t e^{(i+s)r} L_F(s|r)dr \tag{3.10e}$$

And consequently we can have

$$L_F(s|t) = e^{-(i+s)t} + ie^{-(i+s)t} L_D(s) \int_0^t e^{(i+s)r} L_F(s|r)dr \tag{3.11}$$

Then the L.T. of the *effective service time* can be directly obtained via:

$$L_F(s) = \int_0^{+\infty} L_F(s|t) f(t)dt \tag{3.12a}$$

$$= \int_0^{+\infty} e^{-(s+i-iL_D(s))r} f(t)dt \tag{3.12b}$$

$$= L_t(s + i - iL_D(s)) \tag{3.12c}$$

3.2.3.3 PDF of Effective Service Time

The PDF of the *effective service time* can be calculated via the its inverse L.T., which is defined as:[1]

$$f_F(t) = \frac{1}{2\pi i} \int_{\sigma - i\infty}^{\sigma + i\infty} L_F(s)e^{st} ds, \quad t > 0 \tag{3.13}$$

The numerical results of the PDF $f_F(t)$ can be obtained via the trapezoidal approximation [13]:

$$f_F(t) \approx \lim_{\delta \to 0} \frac{2\delta e^{xt}}{\pi} \sum_{n=1}^{+\infty} Re(L_F(x + in\delta))cos(n\delta t) \tag{3.14}$$

where δ is the step size and x is a real number located at right of all singularities.

[1] σ is a real number that located at right of all singularities of $L_F(s)$.

3.2.3.4 M/G/1/K Queue Solution

Considering the V2V data pipe as a server with finite buffer size K, the PDF of the V2V *effective service time* obtained in Eq. (3.14) can be used to solve this M/G/1/K queue status [14].

Defined the 'sampled queue length' as the number of the remaining tasks cached in the queue buffer at the time when a V2V data task has delivered all of the content, which is denoted by Q_i when observed by ith leaving tasks. The 'sampled queue length' at next sampling moment Q_{i+1} can be calculated considering two cases. First, when $Q_i = 0$, i.e., queue buffer is empty at ith moment, then Q_{i+1} depends on the number of new arrivals R_i during the service time of ith V2V task, i.e., $Q_{i+1} = \min\{K-1, R_i\}$. The distribution of R_i can be calculated via:

$$P(R_i = r) = \int_{t=0}^{+\infty} P(r \text{ arrivals in } t \text{ duration}) f_F(t) dt. \tag{3.15}$$

For Poisson arrival tasks with rate λ, we have

$$P(r \text{ arrivals in } t \text{ duration}) = \frac{(\lambda t)^r}{r!} e^{-\lambda t} \tag{3.16}$$

And if $Q_i \neq 0$, then Q_{i+1} equals to $\min\{K-1, R_i + Q_i - 1\}$.

It can be inferred that the series of 'sampled queue length' forms a discrete Markov chain, whose limiting distribution $P(Q_i)$ can be calculated by finding the one-step transition probabilities. When $Q_i = 0$, we have

$$P(Q_{i+1} = q | Q_i = 0) = P(R_i = q), \text{if q} \in [0, K-2] \tag{3.17}$$

$$P(Q_{i+1} = K-1 | Q_i = 0) = \sum_{q=K-1}^{+\infty} P(R_i = q) \tag{3.18}$$

And if $Q_i \neq 0$, we have

$$P(Q_{i+1} = q | Q_i = p) = P(R_i = q - p + 1), \text{if q} \in [p-1, K-2] \tag{3.19}$$

$$P(Q_{i+1} = K-1 | Q_i = p) = \sum_{q=K-p}^{+\infty} P(R_i = q) \tag{3.20}$$

Then together with the normalization equation $\sum_{q=0}^{K-1} P(Q_i = q) = 1$ we can calculate the limiting probabilities of the 'sampled queue length', which can be used to calculate the limiting probabilities of the queue length, denoted by P_k, according to the PASTA property of the Poisson arrival system [14]:

$$P_k = (1 - P_K) * P(Q_i = k) \tag{3.21}$$

where P_K is the probability that the number of the tasks in the queue is K, which is the probability that for any arrived data task, it will be blocked and dropped. We defined the V2V effectiveness as the chances that an arriving data task can be fulfilled via the V2V communication, i.e., all bits of this task can be transmitted, which is denoted by η, then we have

$$\eta = 1 - P_B \tag{3.22}$$

With all values of P_k, we can obtain the mean number of the data tasks in the queue by

$$E(Q) = \sum_{k=0}^{K} k * P_k \tag{3.23}$$

And then the average service delay for a data task can be calculated based on the Little's law consequently.

3.2.4 *Simulation and Discussion*

The proposed V2V queueing framework is validated and the trades off between the V2V data pipe effectiveness and the average time to fulfill a V2V data task are presented in this section. We carry out our simulation in an area around 2500 m × 2000 m, whose road map is from University of Waterloo main campus in Ontario, Canada, as shown in Fig. 3.3a. We generated mobility trace of 200 vehicles by the PTV VISSIM software that randomly move along the map, whose trajectory is plotted in Fig. 3.3b [15].

The parameters of the 'on-period' and the 'off-period' of the V2V contacts are obtained from the moving trace, whose statistics are plotted in Fig. 3.4. Vehicles within the V2V communication range can contact with each other until they drive away. The V2V contacts can also be interrupted by other events, such as blocking LOS [11] by other vehicles or ground facilities, hidden terminal problem, etc., which is assumed to happen randomly during an arbitrary V2V transmission. Specifically, the V2V contact time from the moving trace is counted and fitted to an exponential distribution curve as shown in Fig. 3.4a.We let the vehicle randomly choose one among all available neighbors within communication range to pair for a V2V connection. The statistics of the 'off-period' is shown in Fig. 3.4b, which is fitted to a linear distribution:

$$f_D(t) = 0.02 - \frac{0.02}{100}t, \quad t \in [0, 100] \tag{3.24}$$

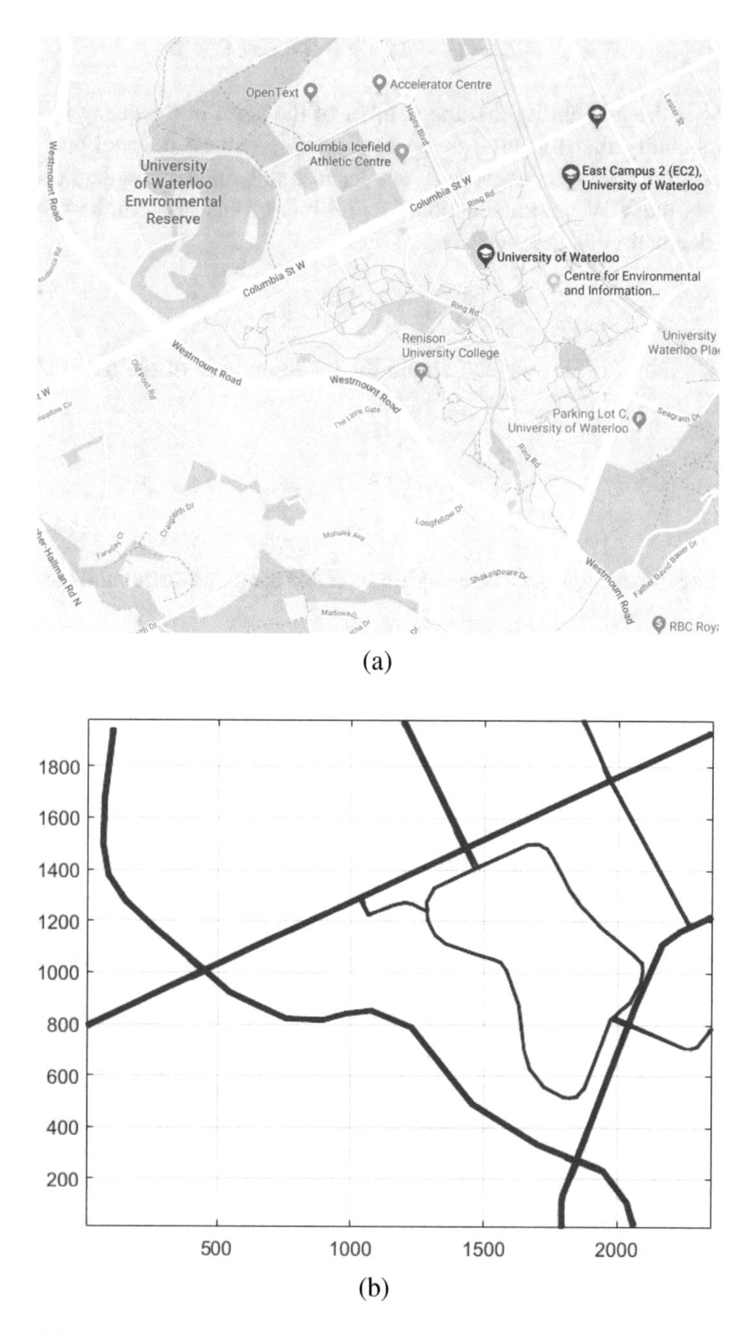

Fig. 3.3 VISIM trace location and vehicle trajectory. (**a**) Map constructed from Waterloo University region. (**b**) Trajectory layout of vehicles

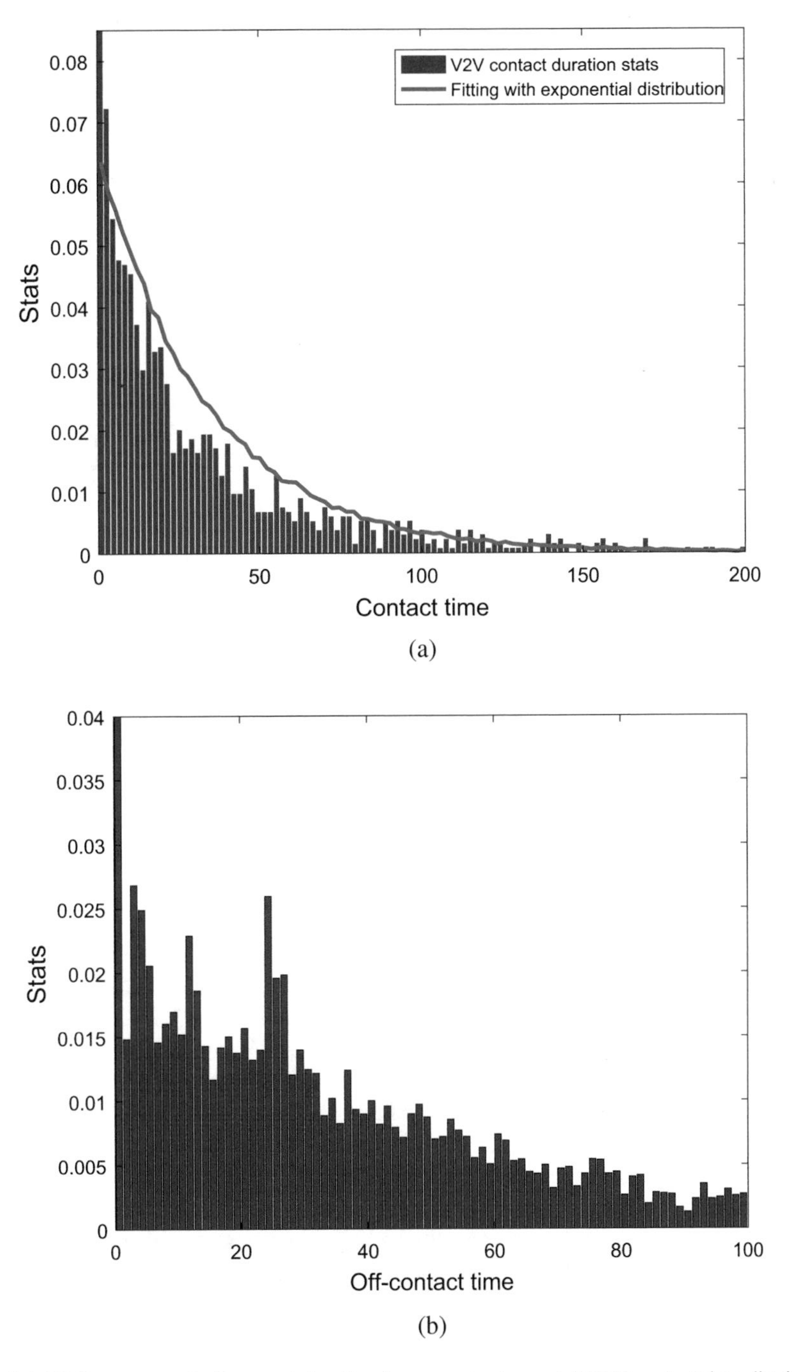

Fig. 3.4 V2V contact and off-contact duration from moving trace. (**a**) V2V contact time distribution. (**b**) V2V off-contact time distribution

Table 3.1 Parameters for V2V contacts and queueing

Parameters	Value
V2V communication range	100 m
V2V average contacts duration	33.82 s
V2V interruption rate i	0.0296
V2V service time t	U[5, 40]
V2V task arrival rate λ	[0.02, 0.03, 0.07]

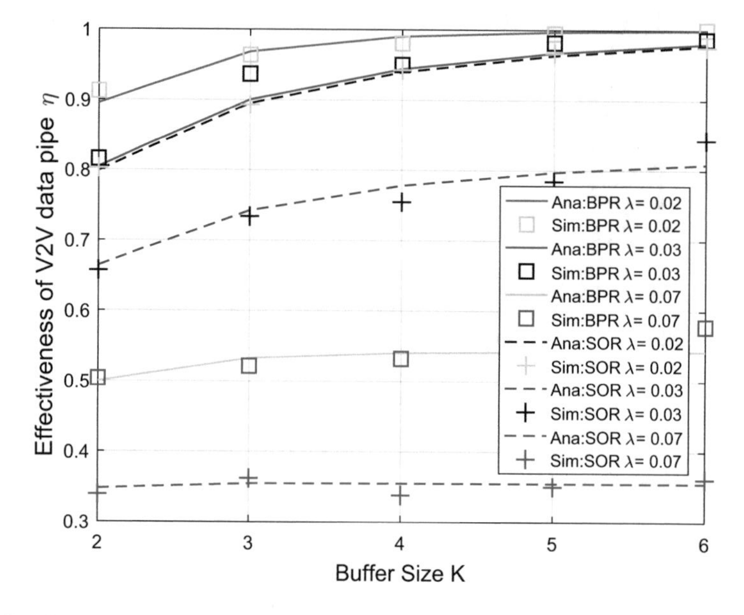

Fig. 3.5 Effectiveness of V2V queueing

We also assume that the time to transmit all contents of a V2V data task (without interruption) is uniformly distributed. And based on the above assumptions and estimations, the V2V queueing parameters are listed in Table 3.1.

Simulation show that our theoretical analysis is consistent with the mobility-trace driven results. Specifically, Fig. 3.5 shows the V2V effectiveness η under various conditions. First, η increases when employing a larger V2V buffer size K, since more data tasks can be cached. And η decreases significantly when the data task arrival rate increases, as the queue is more likely to overflow, especially for the SOR type, which requires the data task to be re-sent when V2V transmission resumes. That's also why the BPR type can achieve higher effectiveness than the SOR type. It is also noticed that continue to increase the buffer size, e.g., lager than 4, may not further increase the V2V effectiveness since the waiting time in the queue would increase significantly and thus the queue will overflow quickly (as supported in Fig. 3.6), and blocking those new arrivals out of the buffer queue.

Figure 3.6 shows the average delay to fulfill a V2V data task, which increases with the size of the buffer K. It can be observed that the delay increases more

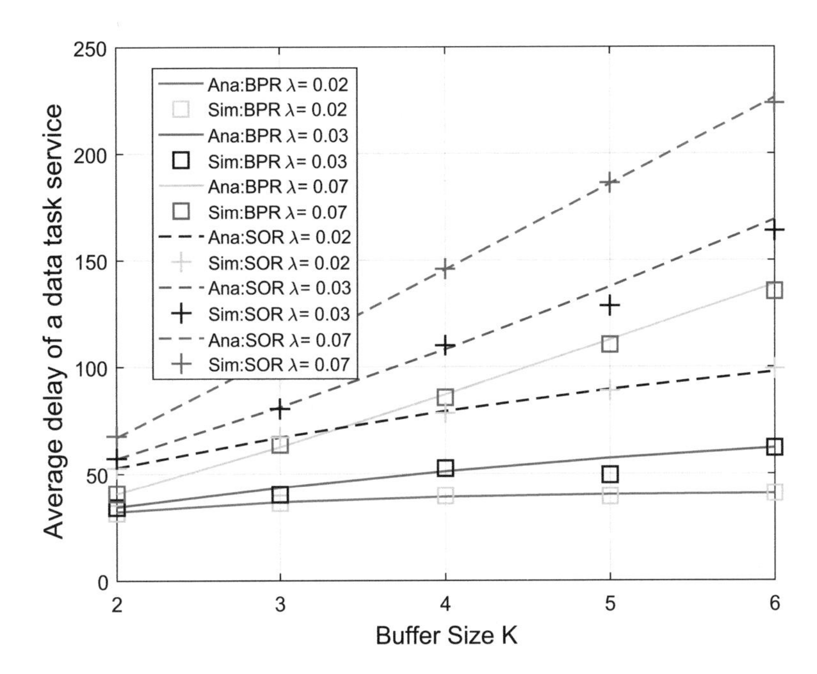

Fig. 3.6 Average V2V service delay

significantly when the arrival rate is high, since more data tasks are in the queue and the average time to stay in the queue for each data tasks becomes much longer. And we can also find that the BPR type consumes much less time than the SOR type, which shows that the time saved via re-transmission from the break point than from beginning is crucial. And when the data task arrival rate increases, the BPR type increase more slowly than the SOR type, as the queue is less likely to overflow that resuming from the beginning bit of a data task would consume much longer time.

The simulation shows how the opportunistic V2V communications can fulfill the V2V data tasks with different arrival rates and buffer sizes. And for the intermittent V2V data pipe server, different data resuming types lead to big difference in terms of the V2V effectiveness and average service delay. The simulation results provide us the basis to extend our framework to other vehicular scenarios, e.g., high way condition, rush hour and low traffic cases, different vehicle mobility pattern, etc., which would present different distribution for both the V2V connection disruption and the 'off-period' duration. The framework can be used to estimate how much data can be transmitted via V2V communication within certain delay, and the trade off between the two in vehicular conditions. Such conclusion can provide useful reference for the V2V communication protocols, e.g., V2V relaying scheme, V2V contend delivering, etc.

3.2.5 Summary

We have proposed a queueing framework to analyze the effectiveness of the opportunistic V2V communication. We have considered the V2V connection disruption and re-establishment, and use an M/G/1/K queueing model with service interruption to describe the V2V transmission for data tasks. We have employed two different data resuming types for a recovered V2V data task, i.e., resuming from break point and resuming from beginning. Simulation have verified our analysis and show the effectiveness and the average service delay of the V2V communication server, as well as the performance difference for two resuming types. Our analysis provides a basis framework that can be extended to more vehicular scenarios, and provide useful information for protocol design and development employing the opportunistic V2V communication.

3.3 Vehicular Offloading via V2X Interworking

3.3.1 System Model

In this section, the ViFi system model is described, which includes network model, vehicle mobility model, Internet access procedure, WiFi offloading queue model and V2V assistance model.

3.3.1.1 Network Model

As shown in Fig. 3.7, assume that all vehicles can access to cellular networks at any moment. WiFi APs are deployed along the roadside and connected to backhaul Internet, and the distances between neighboring APs are the same. There are coverage hole between adjacent WiFi cells, e.g., area $d_3 - d_4$. The ratio of the uncovered area range (d_{uc}) to the covered area range (d_c) is defined as the *uncover ratio*:

$$\eta_u = \frac{d_{uc}}{d_c} = \frac{d_3 d_4}{d_1 d_3}, \quad or \quad \frac{d_6 d_7}{d_4 d_6}, \quad etc. \tag{3.25}$$

The bandwidth of the WiFi AP is shared by all vehicles associated to it, i.e., the average link rate of the tagged vehicle r equals to R/N_a, where R is the overall achievable aggregate rate and N_a is the average number of the tagged vehicle's co-associated vehicles. Besides, vehicles can communicate with neighbors via the V2V communication, e.g., Dedicated Short-Range Communications (DSRC), TV White Space (TVWS) Radio [16]. We assume that the V2V communication can connect

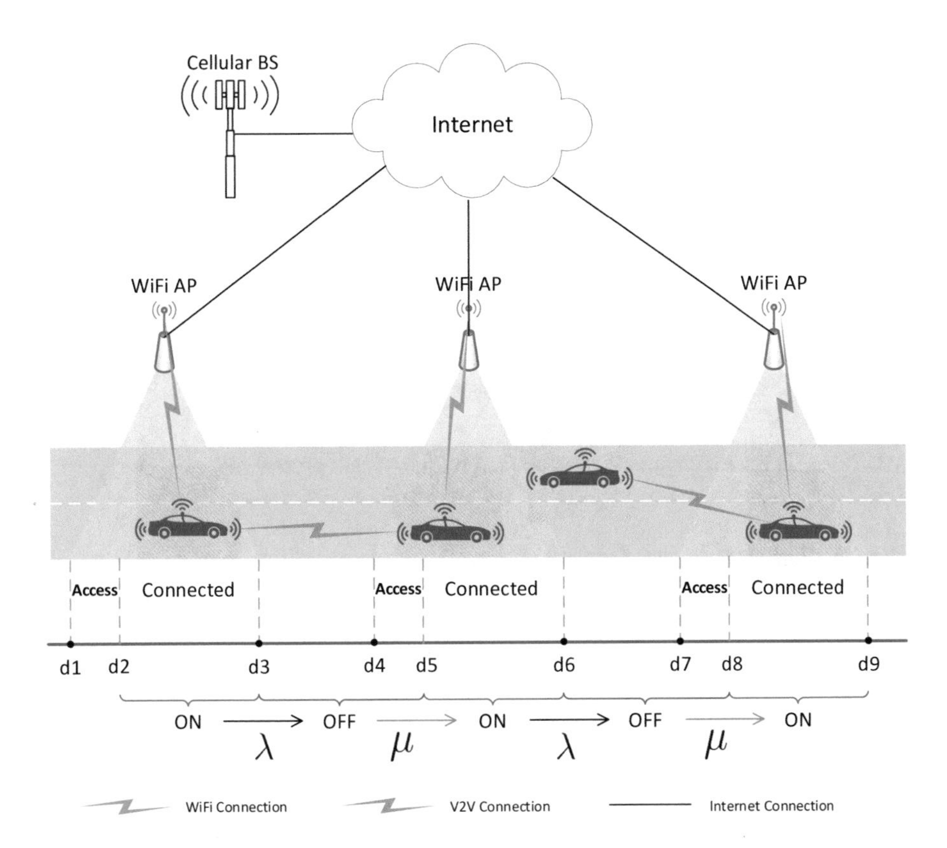

Fig. 3.7 ViFi system model

vehicles in adjacent WiFi cells and the one-hop V2V communication latency can be neglected.

3.3.1.2 Vehicle Mobility Model

We assume that in a long stretch of road, vehicles drive to each lane with Poisson arrival with the parameter of λ_v at velocity of v. In an area of range d, the average number of vehicles equals to $N_l \lambda_v d/v$, where N_l is the number of lanes. Inspired by [9], the switch of the tagged vehicle (the one we consider)'s WiFi connectivity status can be modeled as an *on-off* process. The time duration that the tagged vehicle stays in the connected status can be approximated as an exponential variable with mean $1/\lambda$, while the time duration that the tagged vehicle is off connected with any WiFi AP is approximated as an exponential variable with mean $1/\mu$. And we have:

Table 3.2 Frames exchanged during Internet access procedure

Step	Function	Num. of frames
1.	Beaconning, query	4
2.	EAP-TLS, EAPoL handshake	23
3.	DHCP	2

$$\frac{1}{\lambda} = \frac{d_c}{v} - \tau_{ac},$$
$$\frac{1}{\mu} = \frac{d_{uc}}{v} + \tau_{ac} = \frac{\eta_u d_c}{v} + \tau_{ac}, \tag{3.26}$$

where τ_{ac} is the average delay introduced by the access procedure.

3.3.1.3 Internet Access Procedure

The area the vehicle transverses during the access procedure is defined as the *access area*, which for example, is $d_1 d_2$ (or $d_4 d_5$, etc.) as shown in Fig. 3.7. The access procedure includes the following three steps, in which the management frames exchanged are summarized in Table 3.2.

1. Network detection: In this step, the vehicle will exchange beacon frames or query frames with the AP to obtain necessary information, e.g., radio channel, supported rates, authentication method, etc.
2. alUser authentication: WiFi operators use the authentication to prevent unauthorized users from stealing the WiFi resource, while WiFi users relies on this step to protect their communication privacy. The widely used WPA2-802.1X authentication method is adopted, which requires the handshake between the vehicle and the remote authentication servers.
3. IP address assignment: We use a local Dynamic Host Configuration Protocol (DHCP) server on roadside AP to assign an IP address for associated vehicles.

Besides, generating a management frame requires time for local framing, and sometimes has to wait for the response from remote server (e.g., the TLS handshake frames), which will introduce the 'core delay' before the frame can be exchanged. The core delay during the access procedure used are measured from a real WiFi system.

3.3.1.4 WiFi Offloading Queue Model

Figure 3.8 shows the ViFi queueing model, where the data task request arrival and the departure (means data task is fulfilled) are modeled by an M/G/1/K queueing process, where the time interval between consecutive data tasks is an exponential variable with mean $1/\gamma$. The average time required to fulfill the arrived data task

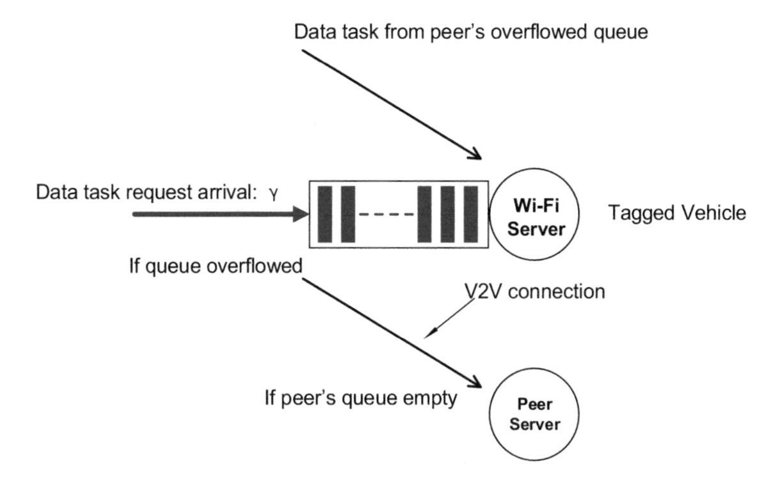

Fig. 3.8 Queueing model of ViFi

equals to $\mathbb{S}/r$, where $\mathbb{S}$ is the average size of the data task and r is the average WiFi link rate of the tagged vehicle, as stated in Sect. 3.3.1.1. We then assume the time needed to transmit all the data of a task follows an exponential distribution with parameter $\lambda_{task} = r/\mathbb{S}$. However, it is possible that during one coverage area, the data task cannot be fulfilled due to limited connection time, thus the serving of the data task has to be deferred until next WiFi AP connection. So that the actual time consumed to fulfill a data task, referred as 'effective service time', follows a general distribution, which is analyzed in Sect. 3.3.3.1.

The offloading performance O_e is evaluated by the ratio of the data tasks transmitted via WiFi APs, which can be calculated by

$$O_e = 1 - P_{B,tag} + P_{B,tag} * P_{assist} \tag{3.27}$$

where $P_{B,tag}$ is the overflowed (or blocking) probability of the WiFi data task queue of the tagged vehicle. And P_{assist} is the probability that the peer vehicle can assist the overflowed data task from tagged vehicle via his own WiFi connection.

3.3.1.5 V2V Assistance Model

When a data task arrives, e.g., to upload an video footage, to download a music file, etc., vehicle will prefer using the WiFi connection to fulfill the data task. If the data task queue overflows, the tagged vehicle will check if his peer can help him to transmit the data through peer's idle resource, otherwise, the tagged vehicle will let the cellular network to fulfill the data task, as shown in Fig. 3.8. The peer vehicle chooses the 'preemptive drop' policy, which allows him to fulfill his own data task with priority. If there is own data task arrival when peer is assisting tagged vehicle's task, then peer will drop the data task from tagged vehicle (which has to be served by cellular network) and start to transmit the data for himself to his own WiFi data pipe. In this way, the peer's WiFi queue should not be affected by the assistance behavior. Since all vehicles' behavior and parameters are the same, the offloading performance can be obtained via the analysis of the tagged vehicle without loss of generality. As the vehicle arrival is Poisson with parameter λ_v, then the probability of the tagged vehicle can find an assistant peer is defined as:

$$P_{v2v} = 1 - P(\text{No accessed peers in neighboring cells}) \\ = 1 - e^{-\lambda_v * 2 * d_c / v} \tag{3.28}$$

Then we have:

$$P_{assist} = P_{v2v} * P_{I,peer} * P_{I,v2v} * P_{na,peer} \tag{3.29}$$

where $P_{I,peer}$ is the probability that peer's data task queue is empty, $P_{I,v2v}$ is the probability that peer is currently not assisting any task from tagged vehicle, $P_{na,peer}$ is the probability that during the assistance of a data task, peer has no new data task arrival.

3.3.2 *Offloading Performance Analysis*

In this section, we present the calculation of the offloading performance, i.e., the traffic ratio that served via WiFi AP, including both self WiFi server and V2V connected peer's AP. We first analyze the access delay, and then derive the effective service time that a data task needs to fulfill through WiFi server. After that we apply an M/G/1/K queueing process to model the status of the data buffer queue, which is used to obtain the offloading performance.

3.3.3 *Access Delay Approximation*

The analysis and experiment in [17] showed that the average access delay τ_{ac} can be estimated as a linear function of the amount of contending nodes, i.e., the number of vehicles associated to the same AP. Thus the average access delay τ_{ac} can be written as:

$$\tau_{ac} = k * N_a + b_0 \tag{3.30}$$

where k and b_0 are constant under certain packet drop rate (PDR, δ) of the management frames during access procedure. The values of k and b_0 used in this article are obtained from the numerical result of [17].

3.3.3.1 Effective Service Time Derivation

To fulfill an arrived data task, it might need transmissions between the tagged vehicle and several APs as the WiFi connection will be interrupted due to the vehicle mobility. Denote $T(x)$ the effective service time of a data task, which requires service time x. x is an exponential variable with mean $1/\lambda_{task}$. Let Φ_λ denote the duration between the beginning moment of the data task service and the upcoming connection interruption, which is exponential with mean $1/\lambda$. Let Φ_μ denote the length of an unconnected duration, which corresponds the duration that the tagged vehicle is out of WiFi connectivity, and is an exponential variable with mean $1/\mu$. And thus:

$$T(x) = \begin{cases} x, & if \quad x \leqslant \Phi_\lambda \\ \Phi_\lambda + \Phi_\mu + T(x - \Phi_\lambda), & if \quad x > \Phi_\lambda \end{cases} \tag{3.31}$$

The Probability Density Function (PDF) of $T(x)$ can be derived using Laplace Transform (L.T.) method, which can be obtained from Eq. (3.31) [12]:

$$\mathscr{T}(s) = \frac{\lambda_{task}}{\lambda_{task} + s + \frac{\lambda * s}{s+\mu}} \tag{3.32}$$

and applying the Fourier-Series (F.S.) method in [13], the inverse L.T. of $\mathscr{T}(s)$, which equals to the PDF of the effective service time, can be approximated via the trapezoidal rules with step size σ:

$$f_T(t) \approx \frac{\sigma e^{\alpha t}}{\pi} + \frac{2\sigma e^{\alpha t}}{\pi} \sum_{m=1}^{+\infty} \mathrm{Re}\left(\mathscr{T}(\alpha + im\sigma)\right) \cos\left(m\sigma t\right) \tag{3.33}$$

where α is a real number located at the right side of all singularities of Eq. (3.32).

3.3.3.2 WiFi Queue Solution

Based on the PDF of the effective service time, we can solve the stable parameters of the M/G/1/K queue to obtain the probabilities to calculate the offloading performance. Denote n_i the number of remaining data tasks in the buffer queue when the ith data task is just fulfilled, then n_i satisfies:

$$n_i = \begin{cases} \min\{K-1, A_i\}, & n_{i-1} = 0 \\ \min\{K-1, A_i + n_{i-1} - 1\}, & n_{i-1} \in \{1, 2, \ldots, K-1\} \end{cases} \tag{3.34}$$

where A_i is the number of the data task arrivals from the beginning moment of the service to the fulfillment of task i, and K is the queue buffer length. From Eq. (3.34), it can be inferred that sampled at every moment a data task finishes, the number of remaining data tasks in the buffer queue forms an embedded Markov chain [14]. By solving the limiting probability of the queue status, we can obtain the blocking probability of the queue $P_{B,tag}$. The probability that k new data task arrivals during the effective service time of ith data task can be obtained by

$$\xi_k = \int_{t=0}^{+\infty} \frac{(\lambda t)^k}{k!} e^{-\lambda t} f_T(t)dt. \tag{3.35}$$

And then the transfer probability matrix of all states P_t can be obtained:

$$P_{t(K \times K)} = \begin{pmatrix} \xi_0 & \xi_1 & \xi_2 & \cdots & \xi_{K-2} & \sum_{i=K-1}^{+\infty} \xi_i \\ \xi_0 & \xi_1 & \xi_2 & \cdots & \xi_{K-2} & \sum_{i=K-1}^{+\infty} \xi_i \\ 0 & \xi_0 & \xi_1 & \cdots & \xi_{K-3} & \sum_{i=K-2}^{+\infty} \xi_i \\ \cdots & \cdots & \cdots & \cdots & \cdots & \cdots \\ 0 & 0 & 0 & 0 & \xi_0 & \sum_{i=1}^{+\infty} \xi_i \end{pmatrix}$$

Thus the steady probability P_k can be calculated together with

$$\sum_{k=0}^{K-1} P_k = 1 \tag{3.36}$$

The probability that the queue length equals to k, i.e., there are k data tasks in the queue, denoted by $\mathscr{P}_k$ equals to

$$\mathscr{P}_k = (1 - P_B)P_k \tag{3.37}$$

Since the real load enter into the queue ρ_r equals

$$\rho_r = \frac{\gamma}{\overline{T}}(1 - P_B) \tag{3.38}$$

where $\overline{T}$ is the average effective service time, which can be obtained from Eq. (3.33). The probability that the buffer length is zero, i.e., no data task in the queue, can be obtained by

$$\mathscr{P}_0 = 1 - \rho_r \tag{3.39}$$

From Eq. (3.37–3.39), we can obtain that:

$$P_{B,tag} = \mathscr{P}_K = 1 - \frac{1}{P_0 + \frac{\gamma}{\overline{T}}} \tag{3.40}$$

The offloading ratio by the WiFi service of the tagged vehicle himself equals to:

$$O_{e,tag} = 1 - P_{B,tag} \tag{3.41}$$

3.3.4 V2V Assistance Analysis

The peer's data task queue can also be solved by the M/G/1/K queueing model following the same method in Sect. 3.3.3.2. And thus,

$$P_{I,peer} = \mathscr{P}_0 \tag{3.42}$$

And the probability that peer is not serving any overflowed task from tagged vehicle can be approximated as

$$P_{I,v2v} = \frac{1}{1 + \gamma P_B \overline{V}} \tag{3.43}$$

where $\overline{V}$ is the average sojourn time the overflowed task from tagged vehicle on peer's server. $\gamma P_B \overline{V}$ is the average number of the overflowed tasks from tagged vehicle during one sojourn time of the previous overflowed task.[2] And $\overline{V}$ can be calculated by

$$\overline{V} = \int_0^{+\infty} \left(\int_0^l t f_\gamma(t) dt + \int_l^\infty l f_\gamma(t) dt \right) f_T(l) dl \tag{3.44}$$

where $f_\gamma(t)$ is the PDF of exponential distribution with the data task arrival rate γ. And the probability that in the assistance duration for an overflowed task from tagged vehicle, peer has no new task arrived, can be calculated by

$$P_{na,peer} = \int_0^{+\infty} \left(\int_r^{+\infty} \gamma e^{-\gamma t} dt \right) f_T(r) dr \tag{3.45}$$

And then the offloading ratio contributed by the V2V assistance can be obtained from Eq. (3.27–3.29) combining the above results.

3.3.4.1 Offloading Ratio Calculation

The overall offloading ratio is given in Eq. (3.27). Besides, the benefit of V2V assistance can be calculated as the *v2v offloading gain*:

[2] Means only one among all the $(1 + \gamma P_B \overline{V})$ tasks can be served if peer is idle.

Table 3.3 Simulation Parameters

Parameter	Denoting	Values
d_c	Range of WiFi coverage area	200 m
N_l	Number of lanes	2
R	Aggregate rate of an AP	54 Mbps
$\mathbb{S}$	Average size of the data arrival	8 MByte
k	In Eq. (3.30): linear function	0.025, 0.052, 0.183
b_0	In Eq. (3.30): linear function	0.47, 0.448, 1.102
σ and α	In Eq. (3.33): step size and a real number	$\sigma = 0.001, \alpha = 0$

$$G_{v2v} = \frac{P_{B,tag} P_{assist}}{1 - P_{B,tag}} \tag{3.46}$$

3.3.5 Simulation and Verification

We conducted simulation based on the parameters listed in Table 3.3. The 'core delay' mentioned in Sect. 3.3.1.3 of each frame is measured from the system described in [17]. The values of k and b_0 from Eq. (3.30) listed in Table 3.3 are obtained when the packet drop rate δ during the access procedure is set to 0.1, 0.5, 0.9 respectively. We run the offloading simulation under various conditions, including different vehicle arrival rate λ_v, data task arrival rate γ, buffer length K, vehicle velocity v and uncover ratio η_u.

Figure 3.9 shows the ViFi offloading performance defined in Eq. (3.27) under different data task arrival rate γ. It can be observed that ViFi can offload the majority of the arrived data task to WiFi APs when the traffic load is low. When γ rises, the offloading ratio decrease since more data tasks will overflow. And by comparing the three cases regarding to different access packet drop rate δ, we can see that the offloading ratio is barely affected until δ becomes a large value. Such conclusion can be used in the access protocol design for vehicular Internet access, especial for high mobility conditions.

Figure 3.10 shows the percentage of the offloading gain through V2V assistance defined in Eq. (3.46). It is shown that when the traffic load is low, e.g., the data task arrival rate γ is small, the V2V assistance can help the tagged vehicle to offload considerable overflowed tasks. When the data arrival rate increase, the peer will be busy to fulfill his own data task according to the 'preemptive drop' policy, and thus contribute less to the V2V assistance. When the vehicle travels with low velocity, given a certain vehicle arrival rate λ_v, the tagged vehicle has a good chance to find an assistant peer in neighboring cells, and thus the offloading gain can be improved.

Figure 3.11 shows the relationship between the ViFi offloading ratio and the uncover ratio η_u. It can be observed that when η_u increases, i.e., the gap areas between two adjacent WiFi cells become larger, the offloading ratio will decrease

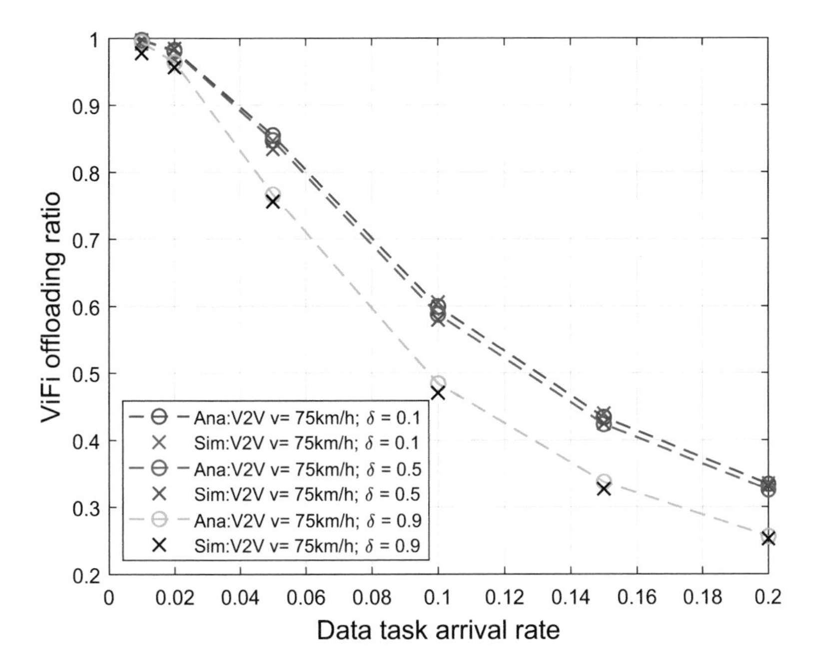

Fig. 3.9 ViFi offloading performance

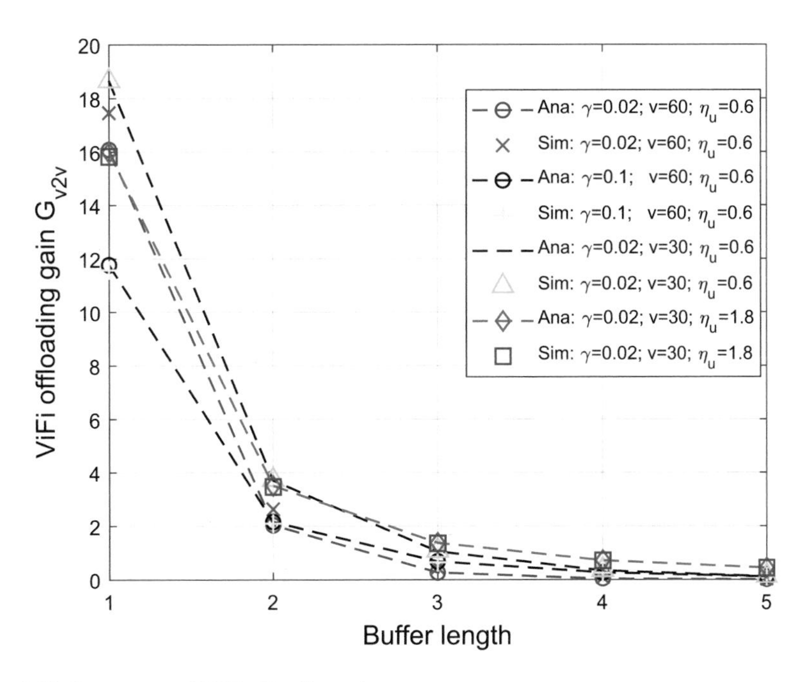

Fig. 3.10 Percentage of V2V offloading gain

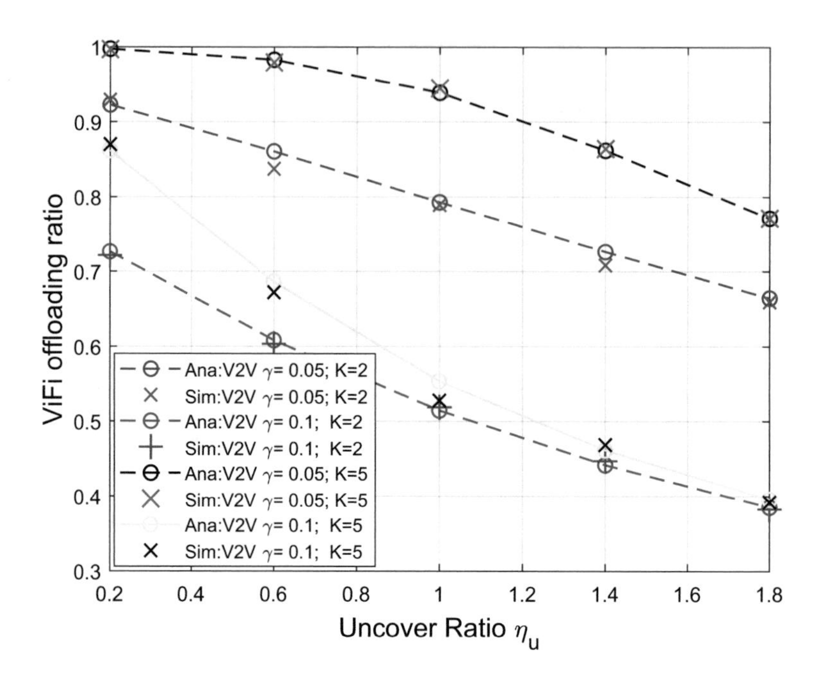

Fig. 3.11 ViFi offloading ratio vs. uncover ratio

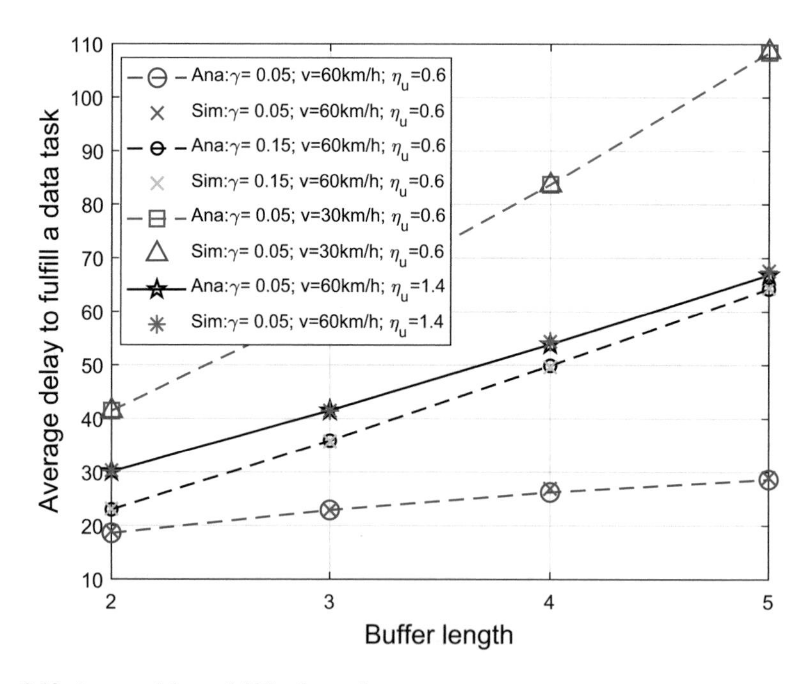

Fig. 3.12 Average delay to fulfill a data task

since the tagged vehicle has to drive through more WiFi cells to fulfill a data task, which leads to longer time for the data task to stay in the queue, and thus makes the queue more likely to overflow. Figure 3.11 also shows that when η_u or the data task arrival rate γ is relatively large, applying a large queue buffer size K can still offload the majority of the data traffic. Such conclusion can be applied in the caching size selection for data delivery in vehicular conditions [18].

However, maintaining a larger K value will increase the number of the data tasks buffered in the queue, and the average duration for a data task stay in the buffer will be extended. And thus, the average delay to fulfill a data task will increase, as shown in Fig. 3.12. The average delay also increases when velocity v, uncover ratio η_u or data task arrival rate γ rises. As a large v and η_u value will let the vehicle have less sojourn duration in a WiFi cell. And when γ increases, the queue length will be extended and the waiting time for a task in the queue will be increased.

3.3.6 Summary

In this section, we have studied the effectiveness of the V2V assistance in WiFi offloading for vehicles. We have utilized the V2V communication to serve as data channel, which can be used to fulfill the data tasks of neighbors, i.e., vehicles can help each other to offload their overflowed data tasks from their local queue. We have applied the M/G/1/K queueing model to analysis the offloading performance of the proposed ViFi scheme with the 'preemptive drop' policy between the tagged vehicle and his peer considering the Internet access overhead. Extensive simulation have been conducted, which show that our analysis is accurate and can provides useful implications for offloading scheme research and design.

References

1. W. Xu, H. Zhou, N. Cheng, F. Lyu, W. Shi, J. Chen, X. Shen, Internet of vehicles in big data era. IEEE/CAA J. Autom. Sin. **5**(1), 19–35 (2018)
2. A. Meola, How the internet of things will transform private and public transportation. http://uk.businessinsider.com/internet-of-things-connected-transportation-2016-10. Accessed 2 Apr 2018
3. N. Cheng, F. Lyu, J. Chen, W. Xu, H. Zhou, S. Zhang, X. Shen, Big data driven vehicular networks. IEEE Netw. **99**, 1–8 (2018)
4. J. Ott, D. Kutscher, Drive-thru Internet: IEEE 802.11 b for automobile users, in *IEEE INFOCOM 2004*, vol. 1
5. R. Mahajan, J. Zahorjan, B. Zill, Understanding WiFi-based connectivity from moving vehicles, in *Proc. ACM SigCom* (2007), pp. 321–326
6. S. Dimatteo, P. Hui, B. Han, V.O. Li, Cellular traffic offloading through wifi networks, in *IEEE 8th International Conference on Mobile Adhoc and Sensor Systems (MASS)* (2011), pp. 192–201

7. B. Han, P. Hui, V. Kumar, M.V. Marathe, G. Pei, A. Srinivasan, Cellular traffic offloading through opportunistic communications: a case study, in *Proc. 5th ACM Workshop on Challenged Networks* (2010), pp. 31–38
8. Y. Wu, J. Chen, L.P. Qian, J. Huang, X. Shen, Energy-aware cooperative traffic offloading via device-to-device cooperations: an analytical approach. IEEE Trans. Mobile Comput. **16**(1), 97–114 (2017)
9. N. Cheng, N. Lu, N. Zhang, X. Shen, J.W. Mark, Opportunistic wifi offloading in vehicular environment: A queueing analysis, in *IEEE Global Communications Conference (GLOBECOM)* (2014), pp. 211–216
10. F. Mezghani, R. Dhaou, M. Nogueira, A.-L. Beylot, Offloading cellular networks through V2V communications—how to select the seed-vehicles? in *IEEE International Conference on Communications (ICC)* (2016), pp. 1–6
11. F. Lyu, H. Zhu, H. Zhou, W. Xu, N. Zhang, M. Li, X. Shen, SS-MAC: A novel time slot-sharing MAC for safety messages broadcasting in VANETs. IEEE Trans. Veh. Technol. **67**(4), 3586–3597 (2018)
12. D. Fiems, T. Maertens, H. Bruneel, Queueing systems with different types of server interruptions. Eur. J. Oper. Res. **188**(3), 838–845 (2008)
13. J. Abate, G.L. Choudhury, W. Whitt, An introduction to numerical transform inversion and its application to probability models, in *Computational Probability* (Springer, Berlin, 2000)
14. S.K. Bose, *An Introduction to Queueing Systems* (Springer, Berlin, 2013)
15. P. Group et al., PTV VISSIM. Retrieved from PTV Group (2015). http://vision-traffic.ptvgroup.com/en-us/products/ptv-vissim/
16. H. Zhou, N. Cheng, N. Lu, L. Gui, D. Zhang, Q. Yu, F. Bai, X. Shen, Whitefi infostation: engineering vehicular media streaming with geolocation database. IEEE J. Sel. Areas Commun. **34**(8), 2260–2274 (2016)
17. W. Xu, H.A. Omar, W. Zhuang, X. Shen, Delay analysis of in-vehicle internet access via on-road WiFi access points. IEEE Access **5**, 2736–2746 (2017)
18. L. Wang, H. Wu, Z. Han, P. Zhang, H.V. Poor, Multi-hop cooperative caching in social IoT using matching theory. IEEE Trans. Wirel. Commun. **17**(4), 2127–2145 (2018)

Chapter 4
Intelligent Link Management for Vehicular Internet Access

In this chapter, we focus on the link layer between a vehicle and an access point. Due to the high mobility of vehicles, the channel condition is highly dynamic due to the varying path loss, shadowing, multi-path fading, etc., which requires an accurate tracking for the link capacity for the proper rate selection for the data packets sent to the air. Traditional model-based link management schemes cannot well follow the various channel variation patterns, and can lead to significantly performance loss especially in high mobility conditions. We propose to utilize multiple model-free ML enabled intelligent schemes that can learn from the previous network trace, which can provide efficient link prediction and near-optimal link rate choice. We analyze different spectrum access for vehicles, including unlicensed bands, TVWS, etc. And more categories of vehicles, e.g., maritime vehicles, flying vehicles, etc., are considered to evaluate the performance gain via the intelligent link management based on machine learning algorithms.

4.1 Background and Motivation

To support the data generated by vehicle users, which is increasing at a tremendous pace, a lot of spectrum resources are utilized to provision the data communication services [1]. WiFi based V2X are more economical while still able to provide considerable performance due to its advantages in cost, deployment and compatibility [2]. And many paradigms have been proposed to utilize the WiFi based IoV, data traffic offloading [3], data caching [4], content delivery [5], etc. In 2012, IEEE published a WLAN based V2X standard, 802.11p, to support the Dedicated Short Range Communication (DSRC) between V2V and V2I working in 5.9 GHz spectrum, which is mandatory for new-manufactured vehicles in north America and has been applied in many cases, such as to support transmitting the safety messages [6], sensing data [7] between different entities on road.

W. Xu et al., *Internet Access in Vehicular Networks*,
https://doi.org/10.1007/978-3-030-88991-3_4

Unlike cellular networks, the IEEE 802.11 standard has not specified the control plane to report the channel status for RA schemes to select the best MCR and leave the RA to user's discretion. Traditional IEEE 802.11 RA schemes can be categorized into two types. One is based on the historical data transmission results [8], the other one is based on the channel sampling, e.g., checking the Channel State Information (CSI) [9]. However, such method can only reflect limited information about the channel variations, e.g., an SNR value only reflects the average signal strength of the inverse channel, and cannot precisely reflect the channel changing, especially for mobile vehicles.

Since the channel capacity itself can be characterized by a lot of trace, including the transmission outcome, signal to noise ratio, etc., in this section, we consider to employ a set of historical SNR values to predict the future channel status or the future rate selection. Specifically, we consider drive-thru Internet that provided by a limited-size WiFi RSU, and vehicular access in a macro-cell from a TV white space (TVWS) base station. Other than vehicles, we also consider other kinds of vehicles, including the unmanned aerial vehicles (UAVs), maritime boats.

In vehicular wireless access, the channel between the vehicle and the base station is affected by many factors, including the path loss that changing with the mobility of vehicles, the slow shadowing, the fast fading or the multi-path fading, etc. Thus it is not easy to setup mathematical model to calculate the relationship between the channel parameters and the proper MCS that should be used. We consider to learn from the previous experience of the vehicles that drive through the same area, and mine the trace data, e.g., the SNR records together with the previous transmission outcome, to predict the future MCS selection by deep learning based classifier and reinforcement learning agent exploitation.

4.2 Reinforcement Learning Based Link Adaptation for Drive-Thru Internet

In this section, we consider to optimize the performance of drive-thru Internet, which is powered by WiFi networks with a great of advantages, such as low cost, quick deployment, no regulation restriction and global compatibility, etc. Despite these advantages, there is a fundamental issue that limit the usage of drive-thru Internet, which is to select the best link rate according to the dynamic channel. Traditional link rate selection schemes are based on statistics of previous transmission results, and are often inefficient in vehicular case, especially that the channel conditions cannot be detected precisely, as shown in Fig. 4.1. In drive-thru Internet, when a vehicle moves toward an AP, the signal strength of the vehicle-AP link first rises, and then falls when the vehicle drives away. We can take advantage of such feature to design our algorithm. In order to improve the channel resource utility, we train an RL agent to choose a link rate for each packet that sent to the air, so that the packet can maximize the channel resource utilization. We utilize the reinforcement learning (RL) to dynamically tune the link rate. RL appears

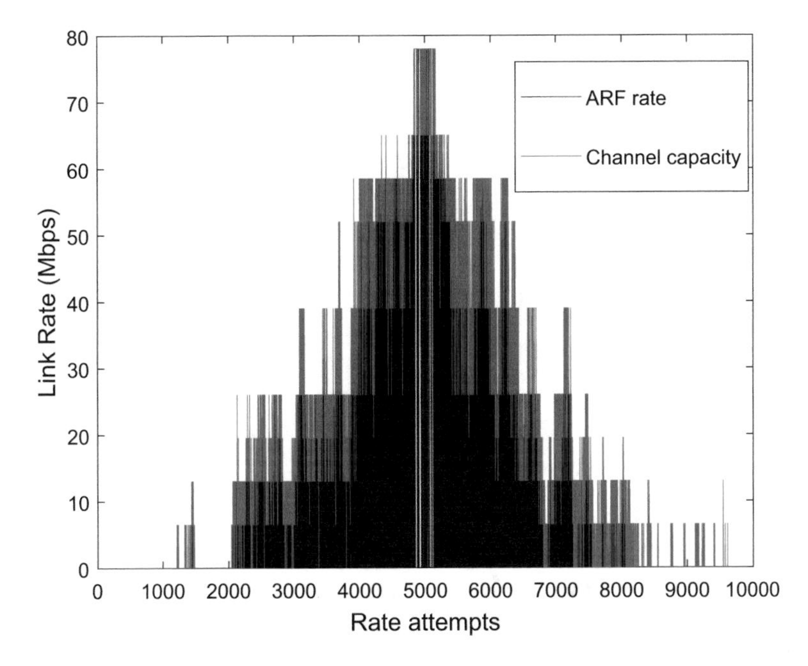

Fig. 4.1 Low ARF RA utilization of channel capacity

to be an effective machine learning tool. By periodical interactions between the agent and the environment, the cumulative reward can be maximized by evolving an optimal decision from the agent. Recently, RL has been used as an emerging technique to address the policy design and strategy determination in communication and networking. Cao et al. adopt machine learning based optimization for joint power control and transmission scheduling, and show that the learning based method can achieve near-optimal performance [10]. GhasemAghaeil et al. present a routing algorithm for wireless sensor networks based on RL, which adapts to dynamic network change and outperforms previous routing protocols [11]. In [12], the authors propose a fast RL method to adjust the transmission policy for the power setting and modulation control to reduce the energy consumption in a point-to-point network. It is shown that RL can lead to the optimal decision via an online learning procedure. The RL agent can exploit the environment, e.g., the network response for different link rate trials. while in drive-thru Internet, the signal strength is determined by the path loss, fast and slow fading. Beside channel based packet loss, the MAC layer also can cause packet drop due to collisions, i.e., two or more packets are sent simultaneously.

We propose an RL based framework for efficient RA in a specific vehicular condition, i.e., drive-thru Internet, which utilizes the historical channel measurements and the transmission feedback to train a neural network (NN) to guide the rate selection for the uplink data transmission for a vehicle. We consider two reasons for transmission failures, i.e., the channel fading and medium access collisions.

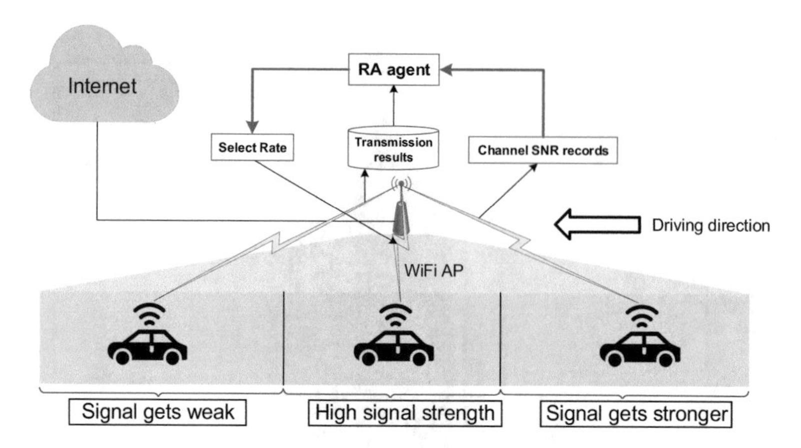

Fig. 4.2 Rate adaptation in drive-thru Internet

Different levels of multi-path fading and medium access contentions are used for the model training and testing. Simulation results show that our RL based RA provides significant performance improvement over traditional RA schemes. The proposed RA is also robust to different levels of channel fading and medium access collisions.

4.2.1 System Model and Problem Formulation

The system model is given in Fig. 4.2. We consider a road where there is an AP to provide the Internet access for drive-by vehicles, who will drive throughput the coverage area with speed v.

4.2.1.1 Network Model

Assume that the vehicle can automatically associate to the roadside AP when it approaches to its coverage area, via protocols such as WPA-802.1X and hotspot 2.0[1] [14].

V2R Channel The wireless channel quality between the vehicle and the AP is determined by three main factors: path loss, shadowing, fast fading.

Data Traffic We assume that the vehicle keeps sending packets of the same size (denoted by D) to the AP, i.e., a saturated uplink traffic condition. Successful transmission of a packet is indicated by the receipt of corresponding acknowledge

[1] Several management packet exchanges are required to finish the user authentication and the IP address assignment in a short time [13].

(ACK) frame. As soon as the vehicle drives into the coverage area, it starts to transmit until it drives out of the area that no packet can be received by the AP.

MAC Protocol IEEE 802.11 distributed coordination function (DCF) is adopted without employing the RTS/CTS scheme. It is possible that multiple vehicles transmit simultaneously, causing transmission collisions. Thus, the failure of a transmission may not only come from channel errors, but also from the transmission collisions in medium access. The collision probability can be estimated by Bianchi [15]:

$$\tau = \frac{2(1-2P)}{(1-2P)(w+1)+Pw(1-(2P)^{m-1})} \tag{4.1a}$$

$$\rho = 1-(1-\tau)^n \tag{4.1b}$$

$$P = 1-(1-\rho)(1-\beta) \tag{4.1c}$$

where τ is the probability that a vehicle transmits in a random time slot, ρ is the probability of transmission collision, P is the probability that a transmission is successful, w and m are the MAC layer parameters of the back off window size and back off count respectively, n is the number of neighbor vehicles that associate to the same AP. The value of the packet loss probability, β, is approximated by applying the fixed rate selection and calculate the average packet loss ratio among all transmission attempts. The MAC collision probability ρ is used to generate the artificial trace in Sect. 4.2.3. Note that packets and frames are used interchangeably.

4.2.1.2 RA Model

The WiFi transceiver on a vehicle selects a rate for a packet to be sent to air. According to the WiFi standard, there are limited discrete rate choices, which depend on the physical layer modulation and coding scheme. All available rates are denoted by $r_1, r_2..r_N \in R$, where R is the set of all. We denote discrete time instants before sending a packet by $0, 1, 2, \ldots t, t+1, \ldots T$, where T is the number of attempts that the transceiver sends the packet, and depends on the sojourn duration of the vehicle and transmission process of each packet. Depending on whether the transmission result I_t is successful (1) or not (0), the time interval between two consecutive discrete time moments $L(t, t+1)$ can be calculated by

$$L(t,t+1) = \begin{cases} \frac{D}{r} + T_{\text{SIFS}} + T_{\text{ACK}} + T_{\text{DIFS}} + T_{\text{BO}}, & I_t = 1 \\ \frac{D}{r} + T_{\text{SIFS}} + T_{\text{DIFS}} + T_{\text{BO}}, & I_t = 0 \end{cases} \tag{4.2}$$

where r is the rate selected for the transmitted packet, T_{SIFS} and T_{DIFS} are the time period of Short Interframe Space (SIFS) and DCF Interframe Space (DIFS), respectively, which are constant; T_{ACK} is the time duration to transmit the ACK frame,[2] T_{BO} is the back off waiting time, which is a random variable depending on the DCF process. For example, if the transmission fails, T_{BO} will increase since the DCF will adopt a larger back off window size for next medium access.

4.2.2 Problem Formulation

The rate selection process can be formulated as a discrete choice problem based on time sequence inputs. The RA agent is required to select a rate, r, from R for transceiver to modulate the egress packet according to the corresponding MCS.

4.2.2.1 Inputs

The input data to the RA agent includes the historical SNR values, e.g., from the received packets, or measured directly from the transceiver. The result of each transmission is recorded to indicate whether the packet with the selected rate can be successfully received. Specifically, the latest H SNR records are used to evaluate the trend of the channel condition. The input data also includes every result I_t of past transmission attempt to train the NN for precision.

4.2.2.2 RA Timeline

Figure 4.3 illustrates the timeline of the rate selection process. When the vehicle needs to transmit a packet to the AP at time t, the RA agent is required to select a rate, r, based on H latest SNR records collected before t. The result of transmission attempt at t is obtained by checking if the corresponding ACK frame is received before the next egress frame transmission at $t + 1$. Both the SNR records and the transmission results are used to determine the rate selection at next time instant $t+1$.

4.2.2.3 Performance Metric

The performance metric of the drive-thru Internet is defined as the overall data amount that a vehicle can transmit to the roadside AP. The objective of the RA is to maximize this metric, which is given by

[2] To improve the ACK frame delivery rate, we set the rate for ACK frames to the lowest rate, 6.5 Mbps. Hence, T_{ACK} is a constant.

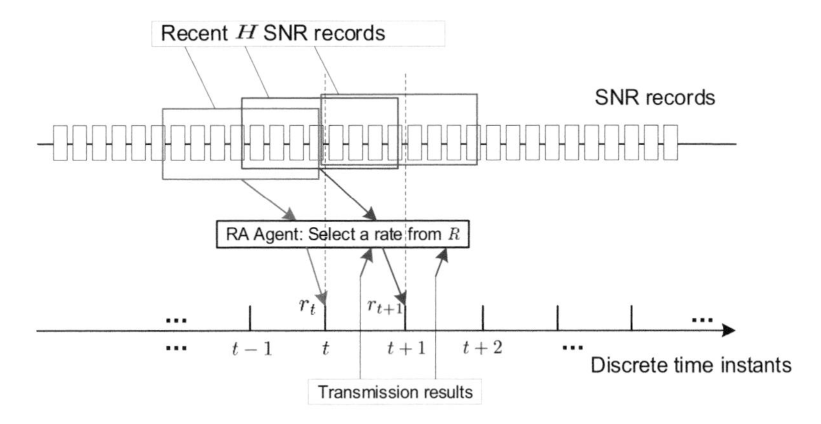

Fig. 4.3 Timeline of the RA process

$$
\begin{aligned}
\max_{r_t, t\in[0,1,2,\ldots,T-1]} \quad & \textstyle\sum_{t\in[0,1,2,\ldots,T-1]} D * I_t \\
\text{s.t.} \quad & \textstyle\sum_{t=0}^{T-1} L(t, t+1) \leq C/v, \\
& r_t \in R,
\end{aligned}
\tag{4.3}
$$

where D is the packet size, v is the vehicle velocity, C is the length of the coverage area, and $L(t, t+1)$ is given from (4.2).

4.2.3 RL Based RA Design

Instead of establishing an explicit relationship between the performance metric and the rate selection based on the input data, the RL based RA adopts a model-free method by training an NN that maps the input data to the rate selection.

4.2.3.1 Network Context Map to RL

The four ingredients that the RA agent deals with in RL are mapped from the following network entities and conditions.

Environment The environment of the RA in drive-thru Internet comprises the network nodes, i.e., the roadside AP and the WiFi terminals including both the vehicle that drives through the coverage area and other WiFi clients. The environment also includes the wireless channel between the vehicle and the AP, which is determined by their distance, shadowing and multi-path fading. When the vehicle sends a packet over the wireless channel, the transmission result is determined together by the channel fading and the MAC layer collisions.

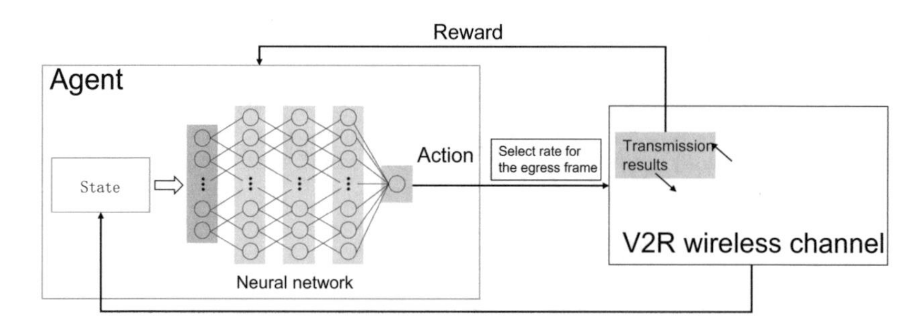

Fig. 4.4 RL structure for RA

State The state of the RA is represented by a set of SNR records from the vehicle's transceiver, i.e., H recent records, $S_t = \{SNR_{M-H+1}, \cdots, SNR_{M-1}, SNR_M,\}$ with SNR_M the most recent SNR record obtained before t. We assume that all SNR records are from the beacon frames periodically broadcasted by the AP per 100 ms.[3]

Action When there is a packet to be sent at time t, the RL agent takes an action, i.e., to select a rate, r_t, from the action space R, which can be obtained from the 'supported rates' section in AP's broadcasted beacon frames. The transceiver then can modulate the signal according to the corresponding MCS of each rate, r_t.

Reward The reward of action r_t depends on the result of the transmission at time t. It is defined as the data amount received by the AP after the transmission attempt:

$$f_R(r_t) = D * I_t. \tag{4.4}$$

If there is no channel error and no medium access collision, the transmission will be successful ($I_t = 1$), the reward is D; Otherwise ($I_t = 0$), the reward is 0.

NN Model The NN model is constructed and shown in Fig. 4.4. At time t, state S_t is mapped to a fully connected layer, which is then connected to three hidden layers before the output layer, whose number of neurons equals to the size of the rate profile R. The output layer gives the Q-values of each rate selection, denoted by $Q(S, r)$, i.e., the scores of action r for a given state S.

4.2.3.2 Learning Structure

The learning structure in Fig. 4.4 shows how the agent learns to take an action of rate selection for the egress packet. At time t, status S_t (i.e., the H SNR records)

[3] The period is the default setting of WiFi devices. The vehicle can also send a probe request frame to the AP, which will then return a probe response frame, to let the vehicle obtain the latest channel SNR.

is input to the NN. The NN then determines the action with the highest Q-value, which is selected and the corresponding rate is applied for the transmission. The transmission result will determine the reward of this action, which is used to train the Q-values in NN. The detail steps of the training and the testing procedures are given in Algorithms 1.

Training Procedure
To maximize the performance metric in (4.3), after the transmission at time t, we use the following metric to update the NN model to improve the long term data throughput:

$$q_t = D * I_t + \max_{r \in R} Q(S_{t+1}, r). \tag{4.5}$$

The NN is updated by adjusting the weights based on the difference between the old Q-value $Q(S_t, a_t)$ and q_t.

The training and testing set comprises the wireless channel data repeatedly sampled by the vehicle in the drive-thru Internet scenarios, and the medium access collision either generated by artificial trace according to Eq. (4.1a)–(4.1c), or by DCF process simulation.

The training procedure includes two simultaneous threads, i.e., the state update thread and the RA thread. The state update thread is to maintain the SNR records from received beacon frames. During the initial phase when the number of collected SNR records is less than H, the default SNR value is set to 0, as shown in thread 1 in steps 5–11. The RA thread is to train the NN model based on the training set. Before sending the packet at time t, the NN model will take action r_t and transmit the packet over the wireless channel. The channel data and medium access collision event from the training set determine the transmission result in step 16 and the corresponding award in step 18. The NN model will be trained to maximize the long-term throughput in (4.3) with q_t. The time from t to $t + 1$ is determined by the transmission result according to (4.3), and the vehicle will move to a new position accordingly. In training procedure, the MAC layer collision probability is artificially generated to train the robustness of the NN model against non-channel noises. The above steps will be repeated every time the vehicle drives through a coverage area.

4.2.3.3 Testing Procedure

Similarly, the testing procedure has the same two simultaneous threads. The state update thread is the same with the training procedure. The RA thread will directly take the action of the NN output and accumulate the data amount transmitted based on the transmission result. By repeating the test procedure based on the testing set, the average throughput that a vehicle can transmit to the AP can be obtained, when drives through the coverage area.

Algorithm 1 NN model training and testing

```
 1: procedure RL BASED RA PROCEDURE
 2: top:
 3:     t ← 0
 4:     h ← 0
 5: thread 1:
 6:     if h < H then SNR[1 : H-h] = 0
 7:     if received AP beacon frame then
 8:         SNR[1 : H − 1] ← SNR[2 : H]
 9:         SNR[H] ← beacon SNR field
10:         h ← h + 1
11:         goto thread 1
12: thread 2:
13:     if associated then
14:         NN ⟸ SNR[1 : H]
15:         r_t ⟸ NN output layer
16:         I_t ⟸ Transmit with r_t
17:         if training then
18:             Reward ← f_R(r_t)
19:             update NN with q_t
20:         else if testing then
21:             Throughout ← Throughout + f_R(r_t)
22:         L(t, t + 1) ← eq. (4.2)
23:         update vehicle position
24:         goto thread 2
```

4.2.4 *Performance Evaluation and Discussion*

To evaluate the performance of the proposed RL based RA, we simulate the drive-thru Internet by constructing the wireless channel, medium access procedure, and vehicle mobility. The PHY and MAC layer parameters are summarized in Table 4.1.

4.2.5 *Experiment Setup*

All network nodes, including the roadside WiFi, the vehicle and co-associated WiFi neighbors, are set to work in 802.11n mode at channel 110 (5.55 GHz). According to the 802.11n standard, the rate profile R is summarized in Table 4.2.

4.2.5.1 Channel Parameters

The V2R channel is simulated with path loss, shadowing and multi-path fading according to the parameters in Table 4.1. The path fading follows the free-space path loss attenuation. The shadowing follows the log-normal distribution. And the multi-

Table 4.1 Parameter values used to generate the channel fading and MAC layer contention

MC parameter	Value	Channel parameter	Value	Other parameter	Value
Backoff window w	16	Power of shadowing	10 dB	Vehicle velocity	20,40,60,80 km/h
Backoff stage m	7	Transmit frequency	5.55 GHz	WiFi coverage	270 m
No. of neighbors n	0,2…18	Multi-path K (dB)	5–25	WiFi power	18 dbm
SIFS	16 μs	Path loss constant/exponent	−12.89 dB/3	Packet size	800 Byte
DIFS	SIFS + 18 μs	Noise power	−90 dbm	WiFi rate profile	802.11n 20MHz & Guard Interval = 800 ns

Table 4.2 MCS index of 802.11n

Rate (Mbps)	Minimum SNR	Received sensitivity (RSSI, dbm)
6.5	2	−82
13	5	−79
19.5	9	−77
26	11	−74
39	15	−70
52	18	−66
58.5	20	−65
65	25	−64
78	29	−59

path fading follows the rice distributed attenuation. The multi-path fading degree K is set to vary from 5 to 25 with step size of 5, which mainly determines the rapid fluctuation of the WiFi signal, as shown in the SNR records in Fig. 4.5. When K is 25 dB, there are less variations than that when K is 10 dB. Figure 4.5 also shows the WiFi signal first increases and then decreases as the vehicle moves along the road. In total there are 200 samples of SNR records for each K, including both the downlink (AP-to-vehicle) and uplink (vehicle-to-AP) channels. Half of the SNR records are used for NN training and the other half are used for testing.

4.2.5.2 Medium Access Trace

To reduce the training time, we use the medium access collision probability ρ in (4.1b) to generate a series of transmission collision events, which is used to determine transmission result I_t in step 16. If a transmission collision happens, the

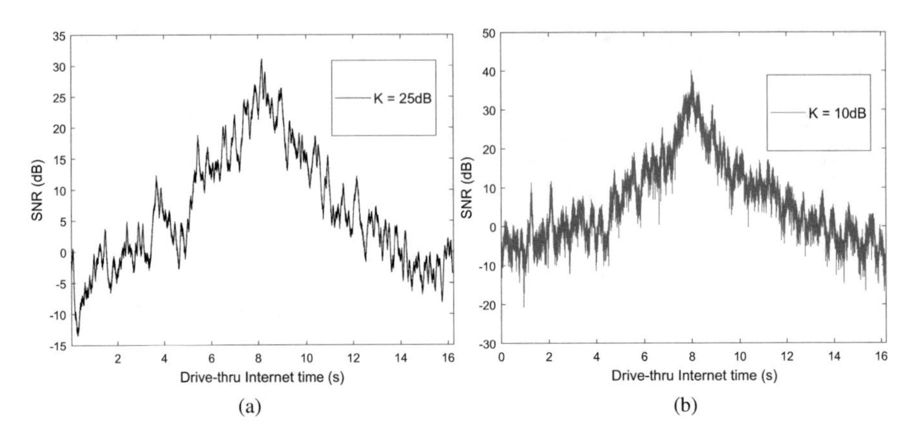

Fig. 4.5 The received signal of the vehicle from the roadside AP. (**a**) RSSI of vehicle for light multi-path fading. (**b**) RSSI of vehicle for heavy multi-path fading

transmission fails even without channel error. In the testing procedure, the IEEE 802.11 DCF medium access is simulated based on the same MAC parameters as shown in Table 4.1. A transmission collision happens when two or more WiFi transceivers transmit simultaneously.

4.2.6 Performance Evaluation

4.2.6.1 Drive-Thru Internet Performance

The throughput performance defined in (4.3) is calculated for each test procedure. The average throughput that the drive-thru Internet can achieve is obtained based on all testing sets. Four RA schemes are tested in the same conditions, whose results are compared with the proposed RL based RA. The optimal throughput is obtained by selecting the highest achievable rate for all packets based on the channel condition in Sect. 4.2.5.1. The ARF and AARF are the traditional RA schemes that widely applied in many WiFi transceivers. The fixed rate selection, as implied by the name, always selects the same rate for every egress packet. In our test, the rate is fixed to 26 Mbps. The random rate selection chooses an arbitrary one in the rate profile in Table 4.5. Figure 4.6 shows the performance comparison when K is set to 10, assuming there is no other co-associated WiFi users. The number of states in the input layer of the NN model is set to 20, i.e., 20 latest SNR records are used to predict the rate selection for the next egress packet. It can be observed that, comparing with the other four RA schemes, the proposed RL based RA can greatly improve the performance of the drive-thru Internet, and achieve almost 83% of the optimal RA throughput.

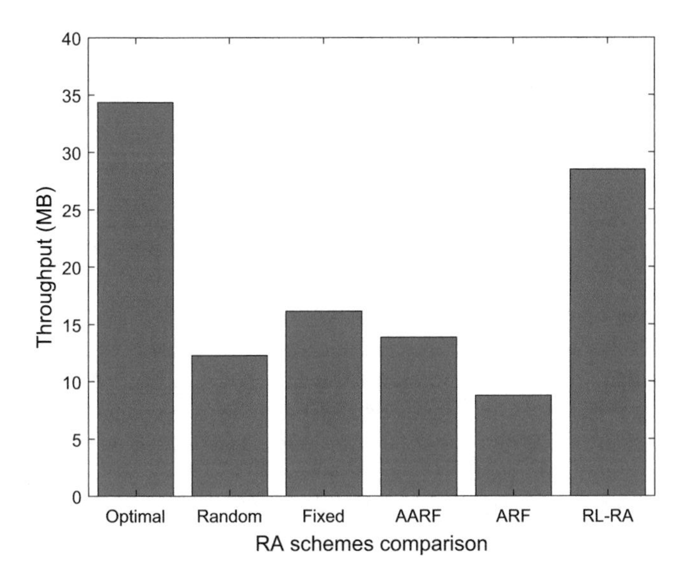

Fig. 4.6 Throughput of different RA schemes in drive-thru Internet

The ARF scheme cannot efficiently adapt to the channel variations in drive-thru Internet. Once there is a transmission failure, ARF would select a lower rate for latter packets. On the other hand, ARF would select a higher rate only if all the previous J transmissions are successful. When the WiFi signal fluctuates quickly due to the multi-path fading, it is almost impossible for ARF to increase the link rate since it is difficult to achieve continuous successful transmissions. When there is a transmission failure, the ARF is too aggressive in attempting lower rates. The AARF scheme can adjust the threshold J to stabilize the rate selection, thus can reduce the interference from quick channel variations, and improve the channel resource utilization. In contrast, the rate selection of the proposed RL RA relies on the recent SNR records and trained memory, which cannot be biased by the transmission failures.

The random rate selection outperforms ARF, and reaches about 88% of AARF. It shows that, in drive-thru Internet, the statistic of transmission results provides poor prediction of channel changes, and even undermines the throughput performance for traditional RAs. The same result can be observed for the fixed rate selection, which does not take into account the transmission results, and thus achieve better performance.

4.2.6.2 Model Generalization

To show the generalization ability of the proposed RA scheme, the trained model is tested, showing strong robustness for different channel conditions and medium access contention levels.

- Robust to multi-path fading: Fig. 4.7a shows the performance for different levels of multi-path channel fading given there are two neighboring WiFi users, which is the throughput results normalized by the optimal value. It can be observed that the performance of ARF RA degrades significantly when K decreases, i.e., when the multi-path fading intensifies, the ARF is sensitive to the short-term fast channel variations, as explained in Sect. 4.2.6.1. The performance of the fixed rate selection is barely affected, without any rate change. The normalized performance of the proposed RA scheme remains roughly the same for all different multi-path fading degrees, around 80% of the optimal performance. We continue to test the proposed RA for different K values from 5 to 25, as shown in Fig. 4.7b. It can be observed that the performance decreases slightly when the multi-path fading becomes more severe, which demonstrates the robustness of the RL based RA to multi-path fading.
- Adaptation to MAC contention: It is difficult for traditional RA to differentiate the reasons of transmission failures, i.e., channel errors or concurrent transmissions. As shown in Fig. 4.8a, when the number of contending vehicles, n, increases, there will be more medium access collisions. As a result, the normalized throughput of the ARF decreases significantly, as ARF attributes all transmission failures to the channel loss and unnecessarily decreases the rate when they are caused by MAC collisions. The fixed rate selection is not affected. The RL based RA scheme shows relatively stable performance, because its rate selection relies on the previous SNR values rather than the statistics of transmission results. In Fig. 4.8b, when n increases from 0 to 18, the RL based RA scheme maintains stable and high performance, which shows its adaptability to different levels of MAC layer contentions.
- Tolerant to velocity variation: We have trained the RL model based on different velocity of vehicles. From Fig. 4.9, it can be observed that when the velocity of the vehicle changes, the proposed RL based RA maintains stable throughput performance. It is because that, when velocity changes, the input SNR series are sampling results from the V2R channel of different sampling rate, which has the same characteristics with each other.

4.2.7 *Feasibility Analysis*

To apply the RL based RA in a drive-thru Internet system, it is worthy to investigate the feasibility of the proposed RA scheme in a real network.

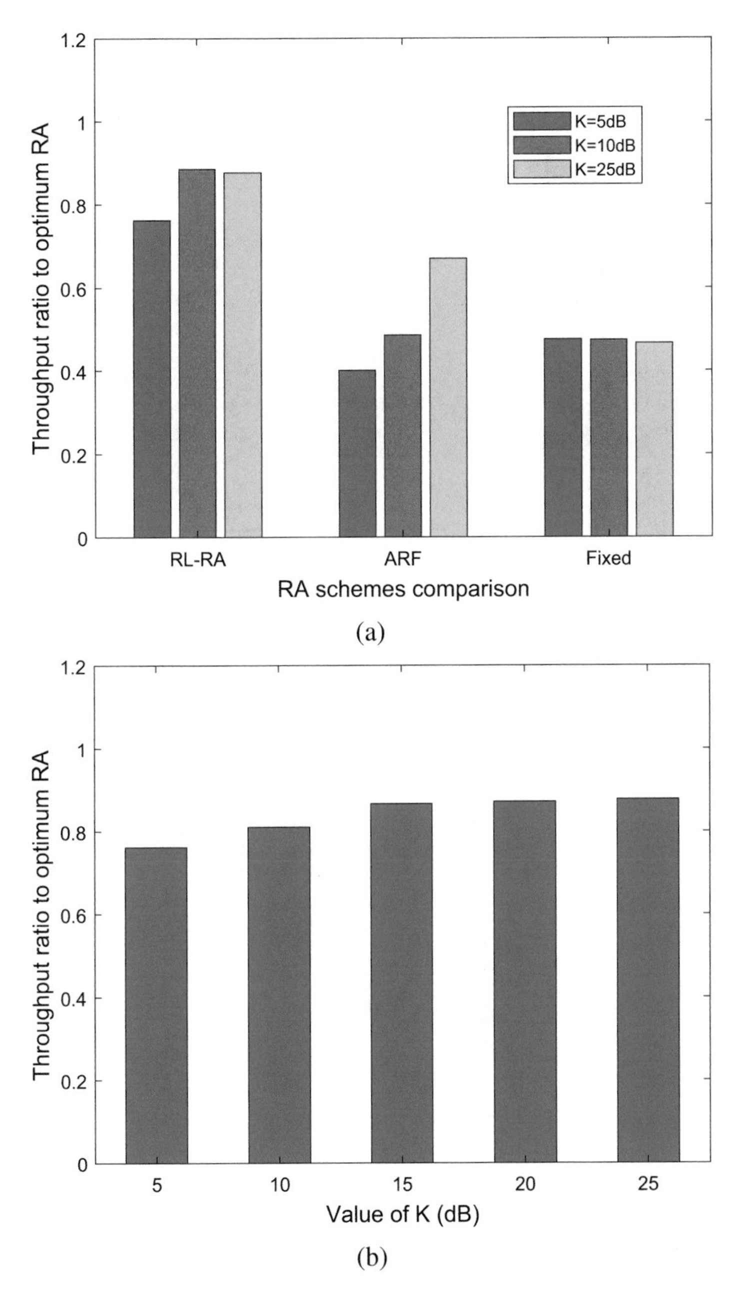

Fig. 4.7 Throughput of RAs for different K values ($n = 2$). (**a**) Throughput comparison. (**b**) Throughput of RL based RA

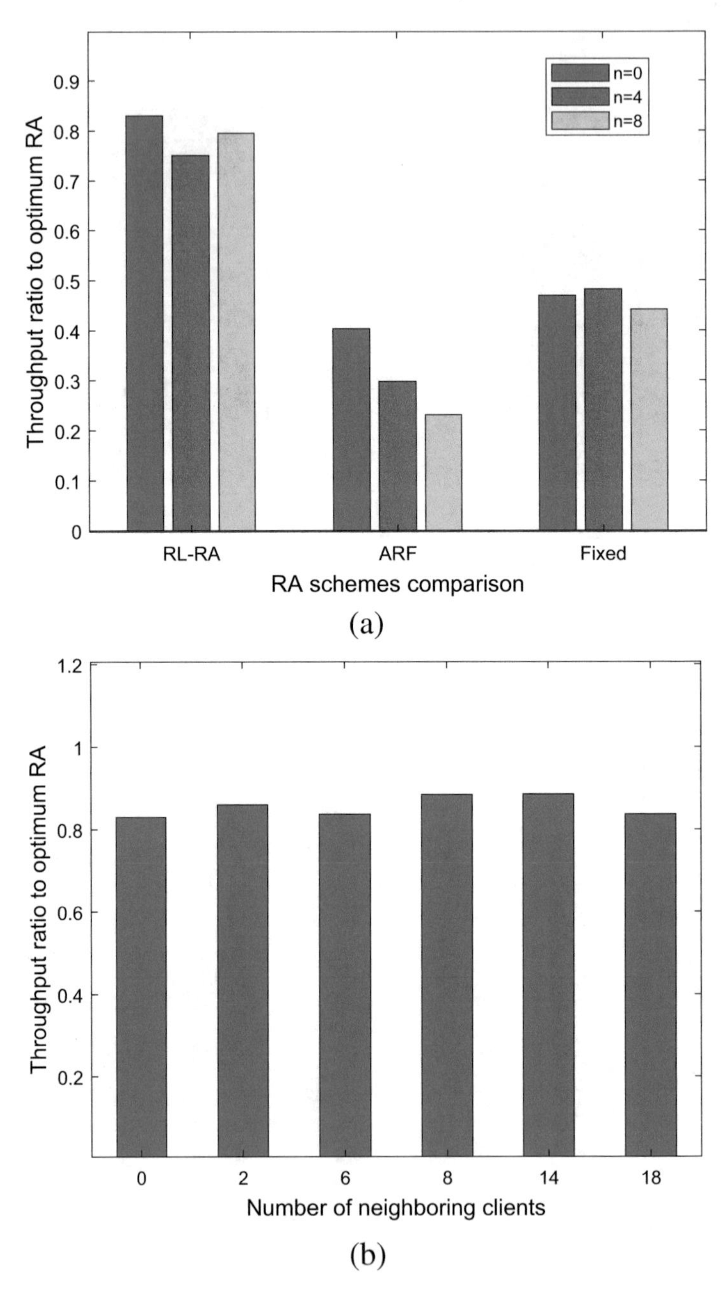

Fig. 4.8 Throughput of RAs for different MAC contention levels ($K = 15$). (**a**) Throughput comparison. (**b**) Throughput of RL based RA

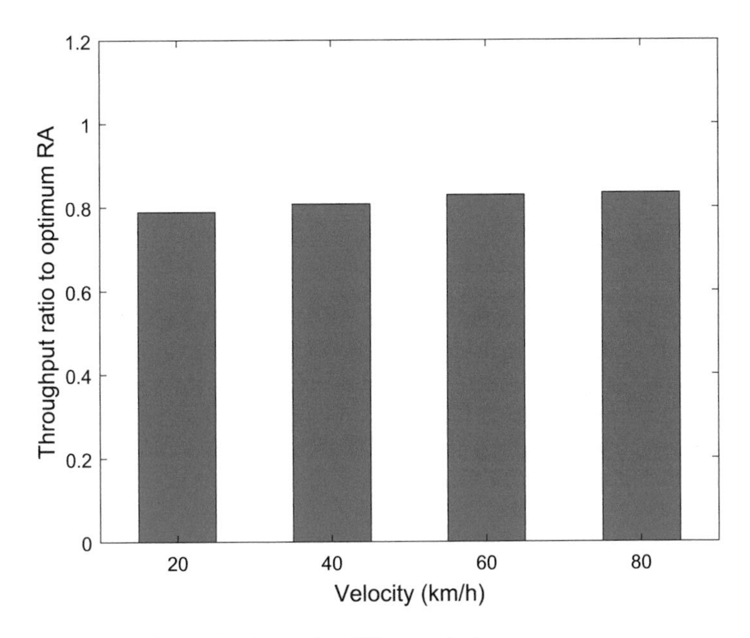

Fig. 4.9 Throughput of RL RA schemes for different velocity

4.2.7.1 Product Compatibility

The IEEE 802.11 standard does not specify the rate adaption, leaving it to users' discretion [16]. For example, the RA scheme is an independent software in most Linux wireless network drivers, which can be customized and reloaded [16]. Hence, the proposed RL based RA does not require modifying the hardware of WiFi transceivers. It can be implemented as software to replace the legacy ones.

4.2.7.2 RL Implementation

The model training may require significant storage space and computing resources. With the development of on-board units, both of their storage and computing capacities are increased significantly, which make it possible to support efficient model training for the proposed RA. The roadside units can also serve as edge computing servers to train the model, and send it to vehicles.

When a vehicle uses the model for rate selection, normally it requires only several hundreds operations of addition and multiplication, which can be finished quickly in many platforms, as shown in Table 4.3. With the growing capability of WiFi transceivers, it is expected that the RL based RA scheme can be implemented on many embedded systems.

Table 4.3 Time consumed for each rate selection using the RL model

CPU	GPU	TPU
Intel Xeon @ 2.20GHz	Tesla K80 (CUDA V.10)	Google TPU
0.387 ms	1.055 ms	0.569 ms

4.2.8 Summary

In this section, we have used the RL tools to deal with the RA problem in drive-thru Internet. Specifically, we use a series of SNR values and transmission feedback to train the RL agent to eventually fit to the channel variation pattern. The model can be well generalized and is robust to various channel conditions and medium access collision. The performance is verified via the experiment, which shows a much improved throughput for vehicles when driving by an roadside AP.

4.3 Deep Learning Classifier Enabled Rate Adaptation for 802.11af TVWS Vehicular Internet Access

TVWS spectrum can be used to provide large area and high throughput coverage. In this section, we propose to use deep learning based rate adaptation (RA) to improve the link throughput of the vehicular wireless communication. The selection problem is modeled as a time series classifier question, which is optimized by several deep learning models that are well trained using the simulated SNR trace. The performance are compared to show that the DL based RA can significantly improve the link efficiency.

4.3.1 System Model

As shown in Fig. 4.10, we consider an area that covered by a TVWS AP and the vehicles are moving along the road in the area. Initially, the vehicle is set to associate to the AP and has obtained the rate list that the AP would support, e.g., by reading the supported rates section from the AP beacon frames. We denote the rate set that both the on-board transceiver and the TVWS AP support by R. The vehicle would generate data packet and send to the TVWS AP periodically. Before sending out a packet, the transceiver should select a rate in R, which is used for the modulation in physical layer.

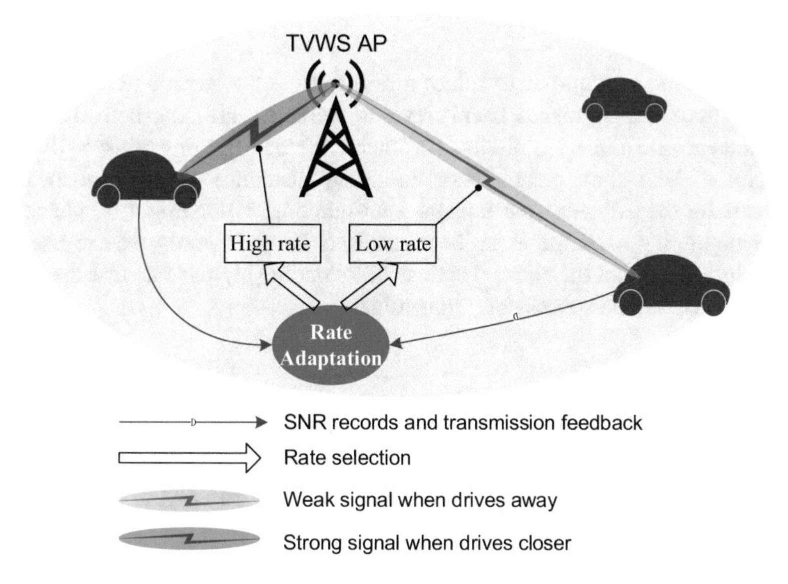

Fig. 4.10 System model of RA in TVWS vehicular access

4.3.1.1 Vehicle Mobility Model

In order to reflect the mobility of vehicles on real roads, the vehicle trajectory is generated on a real map data by the PTV VISSIM software [17]. The vehicle moves along the road lane and would randomly turn in intersection.

4.3.1.2 Channel Model

We assume the uplink and downlink channel between the vehicle and the AP are symmetrical. Both of them are simulated by introducing the path loss, shadowing and the multi-path fading. Specifically, the path loss PL is calculated according to the following equation.

$$PL(d) = 10nlog_{10}d - K_{PL} \tag{4.6}$$

d is the distance between the vehicle and the AP, n is the path loss exponent and K_{PL} is the path loss constant.[4] And the shadowing loss is generated by the log-normal distribution. The multi-path fading follows Rician distribution, whose parameters K indicates the power ratio of the signal of the Line-of-Sight (LoS) to Non-LOS propagation paths.

[4] Determined by carrier frequency.

4.3.1.3 RA Model

The RA scheme is required to select a rate from R for each egress data frame. Our RA model distinguishes from previous works by utilizing both the channel measurement and the transmission feedback. We use the downlink SNR values to evaluate the uplink channel conditions by listening to the beacon frames broadcast by the AP. And the transmission outcome is identified by checking if the corresponding ACK frame has been received. In other words, we use a series of SNR values to predict the channel status and select the highest rate that the channel could support, i.e., be successfully transmitted.

4.3.2 Problem Formulation and DL Solution

The purpose of the rate selection is to maximize the rate of the data frame and expect it can be received by the AP. Instead of calculating single metric to guide the rate selection based on transmission statistics or channel observation, we designed a multi-input single output system that utilize a series of SNR records to make the rate choice for the egress frame. As shown in Fig. 4.11, the vehicle samples the channel by collecting the recent SNR records from the received beacon frames periodically, which is combined as a vector $\mathbb{T}$ of window length w. And then such vector is regarded as the time series input to be classified to several labels corresponding to the rates in R. And thus, such TSC problem can be defined as follows.

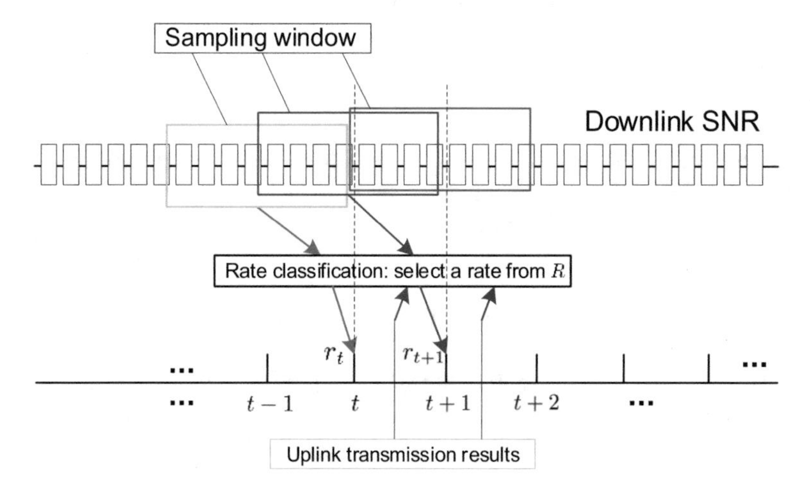

Fig. 4.11 Modeling RA as a TSC problem

4.3.2.1 Requisitions

- A set of all class labels $R = \{r_k\}, \quad k \in [1, K]$. R is the set of all physical rates that both the on-board and the TVWS transceivers supports and K is the number of rates.
- A training set of time series $\mathbb{T}_{train} = \{t^p_{train}\}$, $p \in [1, P]$. The values of all P elements in $\mathbb{T}_{train}$ is the SNRs obtained from the AP beacon frames. The time interval between two consecutive elements is determined by the broadcast cycle of the beacon frames, which is set to 100ms. Each element t^p_{train} is attached with a class label r_k which indicates the rate to be select before the time moments of next element t^{p+1}_{train}.
- A testing set of time series $\mathbb{T}_{test} = \{t^q_{test}\}$, $q \in [1, Q]$. The classifier need to label the time series from t^{q-w+1}_{train} to t^q_{train} to a class r_k, which would be the rate for the egress frame during t^q_{train} to t^{q+1}_{train} moments.[5]

4.3.2.2 Assumptions

- The label of the training set $\mathbb{T}_{train}$ can be obtained by calculating the optimal rate selection based on the foreknowledge of the channel status.
- The classifier only relies on the past time series in TS_{test} as in testing procedure the future channel status remains unknown until the transmission finishes by checking if the corresponding ACk frame has been received.

4.3.2.3 Objective

- In order to improve the utilization of the TVWS channel capacity, our purpose is to maximize the classification accuracy $\mathscr{A}$ on test set, i.e., the ratio of the rate selections that are optimal, which is defined by

$$\mathscr{A} = \frac{\sum_{q=1}^{Q} h_q}{Q},$$

where h_q equals to 1 if the classified label for time series s is the optimal rate, otherwise it equals to 0.

Recent literatures show that such TSC problem can be well solved by applying deep learning models, such as the Multi-Layer Perceptron (MLP), Fully Convolutional Network (FCN) and Residual Network (ResNet) [18], whose structure and hyper-parameters are shown in Fig. 4.12. We investigate the performance of applying these three models to classify the time series of SNR records to predict

[5] If $q - w + 1 < 0$, then set the values of these time series to 0.

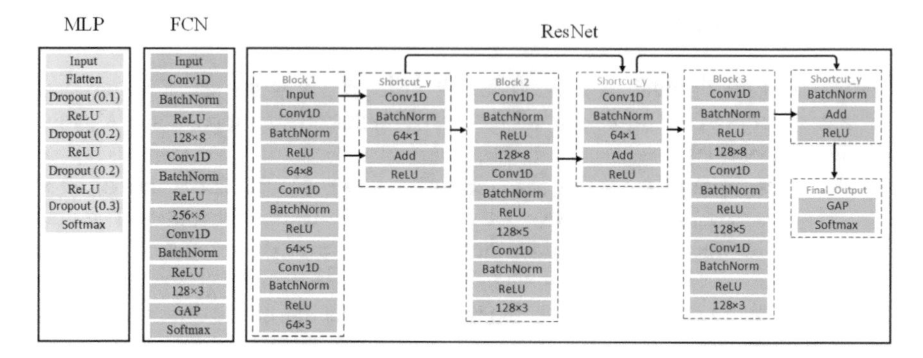

Fig. 4.12 Structure and hyper-parameters of MLP, FCN and ResNet

the optimal rate selection in TVWS vehicular access. Specifically, the MLP model consists of three layers with the adoption of dropout to prevent the co-adaption of neurons. The dropout ratios are set to 0.1, 0.2, 0.2 and 0.3. The FCN model is constructed by three convolution blocks with filter sizes 128, 256 and 128. And a global average pooling layer (denoted by GAP in Fig. 4.12) is employed to significantly reduce the quantity of weights. And the labels are generated by the softmax layer. For the ResNet model, three residual blocks are stacked with shortcut connections for each block to enable the gradient flow through the bottom layer. Similar as FCN, the global average pooling layer and softmax layer are stacked after the main components to produce the final label with reduced number of weights. Moreover, the rectified linear unit (ReLU) is chosen as the activation function for all three deep learning models to prevent the so-called gradient saturation.

4.3.3 Evaluation of the TSC for TVWS RA

In this section, we evaluate the performance of three deep learning model based TSC solutions compared with the conventional RA schemes.

The experiment is carried out in an area around 2000×2000 m with the map data from the main campus of University of Waterloo in Ontario, Canada, as shown in Fig. 4.13.

A TVWS AP is setup in the middle of the area, both of the on-board transceiver and the AP are working on 650 MHz. The V2I channel is simulated by adding the path loss, shadowing and multi-path fading according to the parameters in Table 4.4.

The rate set that can be labeled is shown in the MCS index Table 4.5 according to the measurement in [19]. If the SNR is smaller than the minimum required value, then the transmission would fail, i.e., the h_q in Eq. (4.7) equals to 0. While if the SNR is larger than the minimum one, then the transmission would success and $h_q = 1$.

Fig. 4.13 Campus of University of Waterloo

Table 4.4 MCS index of 802.11af

Parameters	Value
Path loss exponent n	3
Path loss constant K_{PL}	−12.89 dB
Noise power	−90 dB
Transmitting power	30 dbm
Transmitting frequency	650 MHz
Noise power	−90 dbm
Shadowing power	10 dB
Multi-path K	10 dB

Table 4.5 MCS index of 802.11af

Rate (Mbps)	Minimum SNR	Modulation
2.4	2	BPSK
4.8	5	QPSK
7.2	9	QPSK
9.6	11	16-QAM
14.4	15	16-QAM
19.2	18	64-QAM
21.6	20	64-QAM
24	25	64-QAM
28.8	29	256-QAM

We generated 292 vehicle trajectories by the PTV VISSIM software that moves along the road sketch from the real map in Fig. 4.13. Each vehicle would record the SNR value every 100 ms to form a time series, whose total time duration is around 1000 s. Among the total 292 vehicles, the time series SNR records of 146 vehicles are used to form the training set $\mathbb{T}_{train}$, while the remaining ones forms the testing set $\mathbb{T}_{test}$.

The value of the sampling window w is set to from 5 to 50 with step size of 5. And thus the train and test process are conducted for 10 runs to compare the impact of the w to the RA performance. The learning rate is denoted by L_R. For FCN and ResNet L_R is set to 0.001, while for MLP, L_R is set to 1 due to its relatively small number of tunable parameters. The batch size of all the three models are set to 1024. All DL based training and testing are conducted on Google Collaboratory by using Keras with Tensorflow as the backend.

4.3.4 Performance Evaluation

The performance of the classify accuracy is obtained by averaging the results from 10 iterations and up to 100 epochs on each w value. During the training process, the actual epoch is taken when the loss value reaches its lowest value.

4.3.4.1 Classification Accuracy

The accuracy of different DL models, which reflects the ability of the RA to predict the optimal rate selection, are shown in Figs. 4.14a, 4.15a, and 4.16a as the box plots with minimum, maximum, 25 percentile accuracy, median accuracy, 75 percentile accuracy, and above the box plots the average accuracy in blue. It can be seen from Figs. 4.14b, 4.15b, and 4.16b that the ResNet model outperforms the other two with the best classification accuracy of 76%. And it can also be observed that almost all trained models provide stable accuracy performance for different w values, which show that the three classification models are robust to different training process on the given TS_{train} and TS_{test}. It can also be observed that increasing the sampling window size w does not necessarily improve the accuracy. When w is small, increasing w could include more history information, thus can better predict the future channel condition, as seen for all the three models, while when w is large enough, i.e., more than 30, increasing the value of w could no longer contribute to the accuracy, as too old SNRs provide little information on future channel conditions in TVWS vehicular access, and sometimes would even deteriorate the accuracy performance as shown in Fig. 4.15a.

Furthermore, the training processes of MLP, FCN and ResNet models are shown in Figs. 4.14b, 4.15b, and 4.16b, indicating the epoch moment when the highest accuracy achieved. It can be observed from Fig. 4.14b that the loss of the MLP model fluctuates drastically and fall slowly. Figure 4.15b shows that the FCN model

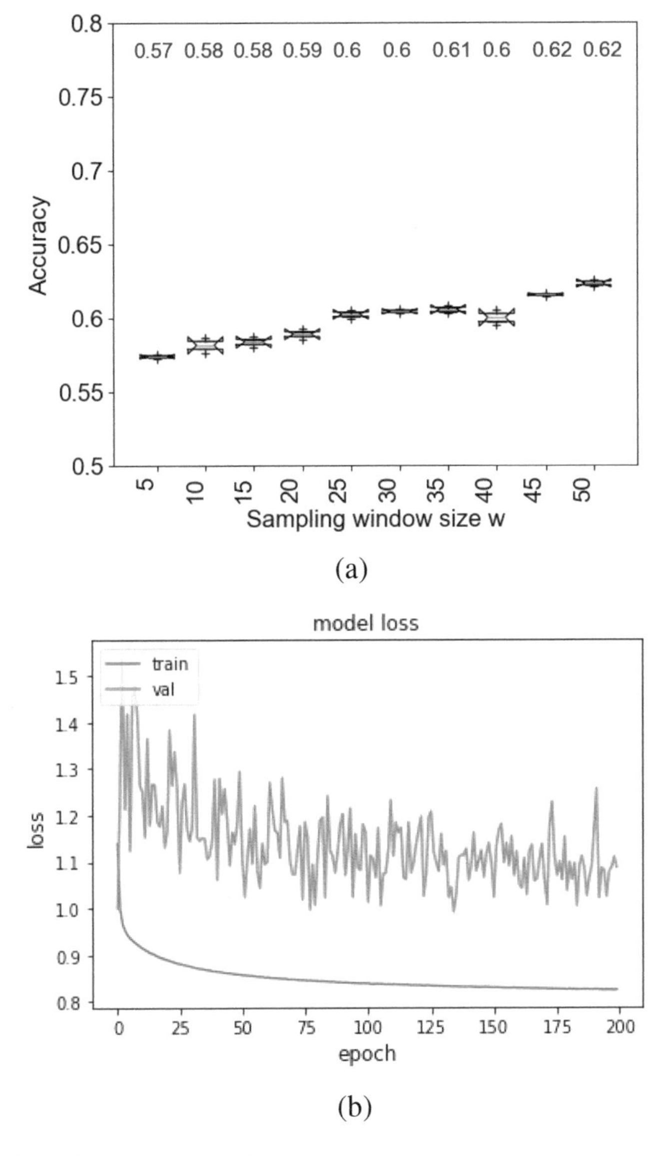

Fig. 4.14 MLP training and results. (**a**) Accuracy vs. w. (**b**) Loss vs. epoch

could converge efficiently at earlier epoch. And Fig. 4.16b shows that the ResNet model is over-fitting since the loss would increase rapidly after epoch 20. The reason comes from the fact that the ResNet model to too complex and having more tunable parameters than MLP and FCN models.

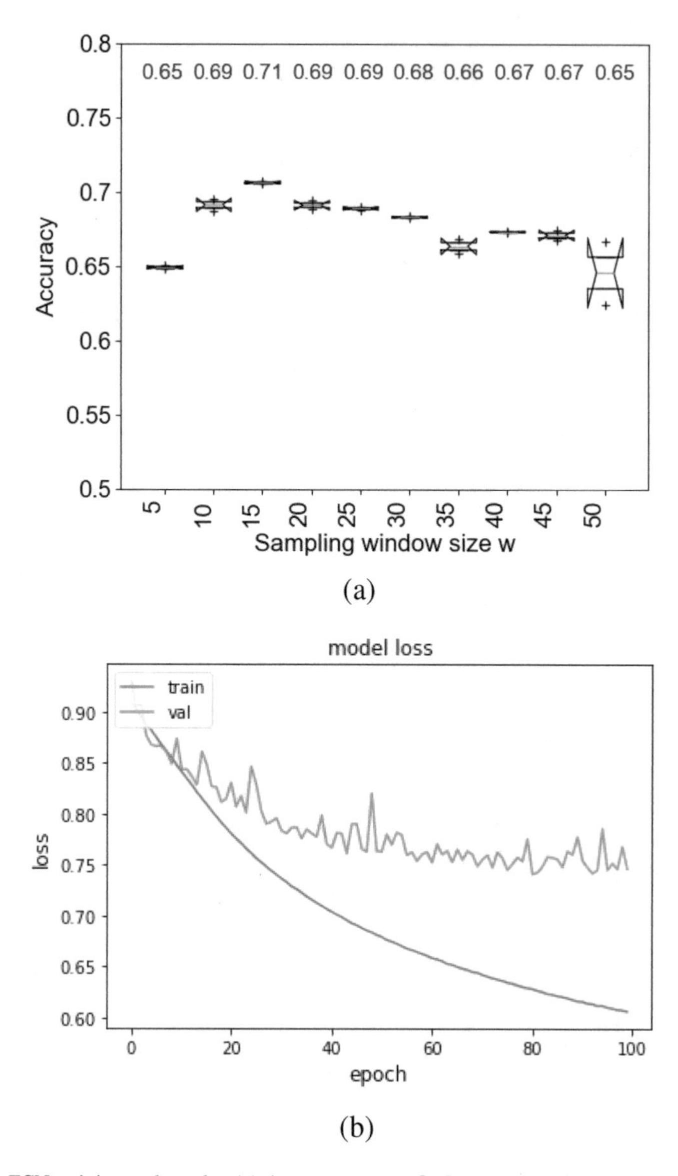

Fig. 4.15 FCN training and results. (**a**) Accuracy vs. w. (**b**) Loss vs. epoch

4.3.4.2 Throughput

The throughput of the vehicle is defined as the overall data amount that the vehicle can send to the TVWS AP. To evaluate the performance of different RAs, we introduce the 'normalized throughput', T_N, defined by dividing the throughput of each RA T_{RA} to the optimal selection result T_{opt}, i.e., $T_N = T_{RA}/T_{opt}$. The

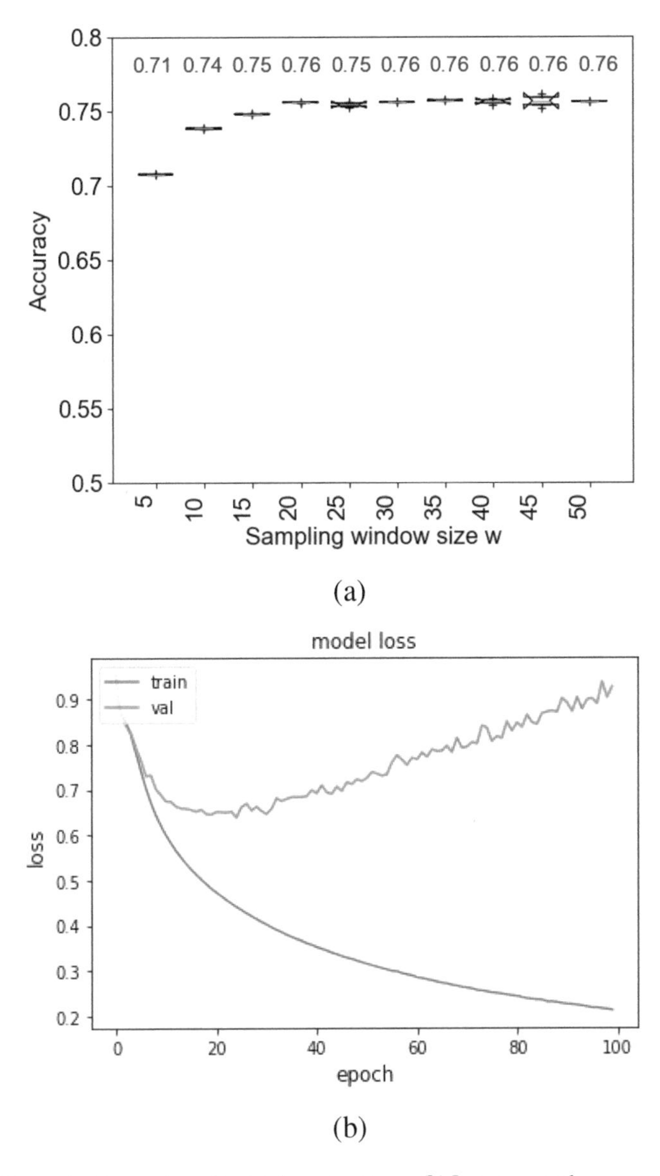

Fig. 4.16 Resnet training and results. (**a**) Accuracy vs. w. (**b**) Loss vs. epoch

comparison of the three DL model based RAs and conventional RAs are shown in Fig. 4.17, which show that the DL based RA could almost improve the overall throughput significantly. It also show that even the fixed rate selection (fixed to 14.4 Mbps) and random RA are better than the traditional ARF scheme.

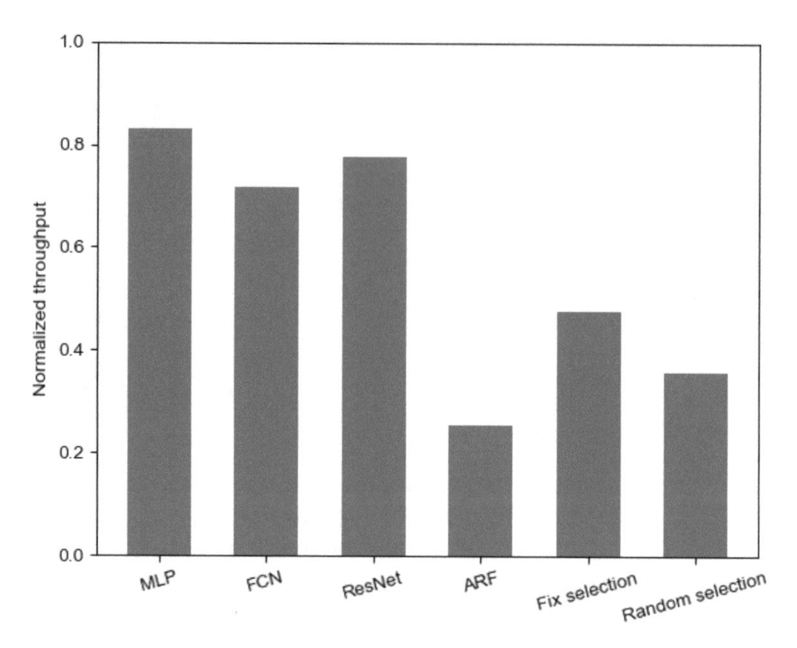

Fig. 4.17 Modeling RA as a TSC problem

4.3.5 Summary

In this section, we have utilized a classifier powered by three deep learning model to categorize the proper rate selection for the next egress frames. The experiment in our simulation environment, which are superposed by path loss, slow and fast fading effect, has shown that the rate selection can be improved by 76%, and can significantly improve the throughput. The training process shows the epochs to converge the ML model, and FCN model are the most efficient one with less sampling window size. The analytical methods and proposed methodology can provide inspiring guidance for applying DL models in mobile communication.

4.4 Autonomous Rate Control for More Categories of Vehicles

In this section, we consider more general cases, where different categories of vehicles are considered to optimize the MCS selection for their wireless link with the base station. We consider to train an RL agent to exploit the channel variation pattern, which is mainly determined by the mobility pattern. We consider two mobility cases, e.g., the vehicles drives in a block streets and a UAV cruising in a circle. In this case, vehicles are always moving in same trajectory, and such feature can help to predict the channel status, as the signal strength variations usually have the same pattern. In this section, we seek to exploit such pattern using

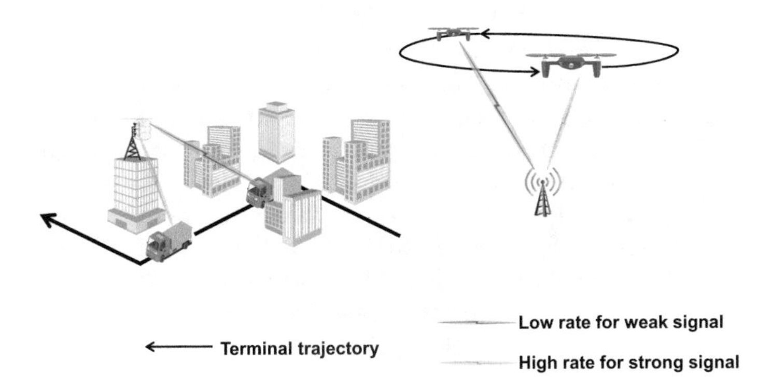

Fig. 4.18 System model of RC

a deep reinforcement learning (DRL) agent to autonomously control the link rate to maximize the link utilization by designing a dedicated reward function for each iteration. The train agent is shown to have good generalization capability under different conditions.

4.4.1 System Model

As shown in Fig. 4.18, we consider wireless access for connected vehicles and UAVs whose mobility are similar, i.e., traveling along fixed trajectories. We utilize such important characteristics to optimize the communication performance.

4.4.1.1 Network Model

We focus on a wireless network including one target vehicle, and assume that it can access to the channel periodically, i.e., the vehicle is allowed to transmit a packet to the channel once per t_a units of time. The result of a transmission attempt is determined by the channel condition and the rate selected for the transmitted frame. The terminal is also assumed to be backlogged, i.e., always has packet to send in the transmission buffer queue.

The signal-to-noise-ratio (SNR) is used to characterize the channel condition, which is determined by the transmission power and the channel loss, including the path loss, shadowing and multi-path fading in mobile conditions [20]. The path loss is calculated as follows.

$$PL(d) = 10nlog_{10}d - C_{PL} \tag{4.7}$$

n is the path loss exponent, and C_{PL} is a constant value that is determined by the wireless frequency, gains of the transmitting and receiving antennas. The path loss contributes the large-scale signal attenuation, and is mainly determined by the distance between the terminal and the access station.[6] As the vehicle often travel along fixed route, the path loss would have similar variation pattern, which can be recognized by a trained neural network and predict the trend of the signal strength.

Shadowing (also called the slow fading) is caused by obstacles in the signal path. It can be caused from buildings, road facilities, etc. When the terminal moves within the same region, the shadowing effect should be similar and varies over a large-scale in distance. We assume that the shadowing follows the lognormal distribution [21].

The multi-path fading comes from the reflection and scattering that lead to multiple signals arriving at the receiving antenna. We assume that the signal power from line-of-sight (LOS) propagation path contributes the major part of the combination of multi-path components, which is assumed to follow the Rice distribution. The ratio of power contributed by the LOS path to the sum of the remaining paths is defined as the shape parameter K. To evaluate the generalization capability of the learning model, we consider different values of K to present different multi-path fading levels.

4.4.1.2 Mobility Model

We consider vehicle that repeatedly moving along certain trajectories, i.e., vehicle move along streets, UAV travel along a circle. Specifically, the vehicle terminal and the UAV would repeatedly travel along the same routes, which is common in daily lives. For example, vehicles drives through the same path from one location to another. And UAVs are hovering with same motion parameters to provide similar surveillance services. Figure 4.19 shows the typical motions for vehicle terminal and UAV terminal that being considered. The vehicles are moving along the street roads following the velocity limits. Both standard square blocks streets and non-standard square[7] region are considered. And the UAVs are cruising along a circle in the air. The distance between UAV and the ground station is determined by the UAV's altitude h, the horizontal distance ℓ and the radius of the circle r. The velocities of the ground vehicle and the air UAV are denoted by v_g and v_a respectively.

4.4.1.3 Rate Profile

Before actually sending the packet to air, the transceiver of the vehicle should select one MCS from the rate profile R, and then the data bits are mapped to corresponding electromagnetic signal. The link capacity, i.e., the achievable rate,

[6] The higher the mobility, the lower the coherence time of the channel.

[7] As shown in Fig. 4.21a.

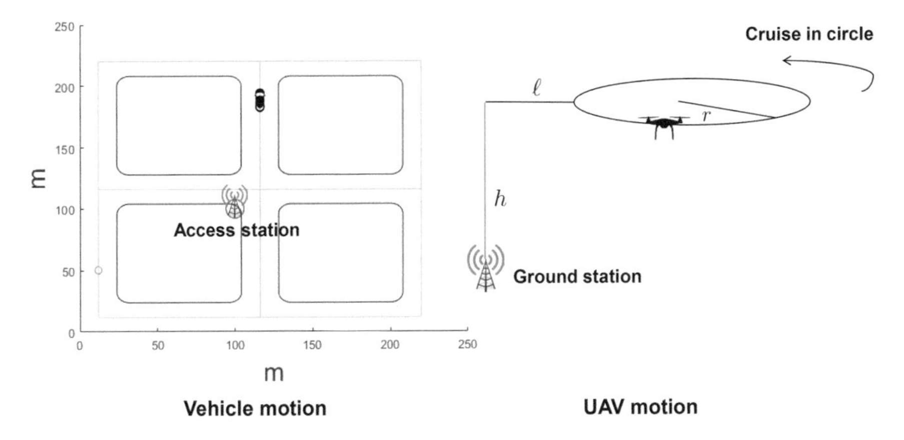

Fig. 4.19 Mobility pattern: vehicle and UAV

is highly dependent on the SNR level. As many literatures and experiments have investigated, we assume that the minimum achievable rate is a step function of the link SNR [22]. To simplify the analysis, we assume that a successful transmission attempt should satisfy the following two conditions together.

- The received signal strength indicator (RSSI) is larger than the minimum input level sensitivity;
- The selected rate is less or equal to the maximum rate the link can support with current SNR value.

4.4.2 Problem Formulation and DRL Based RC

4.4.2.1 Problem Objective

As shown in Fig. 4.19, when the vehicle travels, the distance between the terminal and the access station changes, the majority of the channel attenuation, i.e., path loss component, changes correspondingly. The shadowing and multi-path fading are also highly correlative with the location of the terminal. We focus on the uplink communication that data traffic is required to be sent from vehicle to access station. We consider L levels of MCS, i.e., there are L rates in R. The terminal can collect discrete channel SNR samplings from the received beaconing frames periodically broadcast by the access station.[8] The objective of the problem is to maximize the average channel utility, which can be defined as follows:

---

[8] For example, the IEEE 802.11 terminal can read the RSSI and SNR from both the beaconing and data frames.

$$\max_{r_i} \frac{1}{N} \sum_{i \in [1,2,\ldots,N]} z_i * r_i \\ r_i \in R, \tag{4.8}$$

where N is the total number of packets that the transceiver attempted to transmit. z_i is the result of the ith transmission attempt. It is set to 1 if the transmission is successful, and to 0 otherwise. z_i can be obtained based on reception of the corresponding ACK frame.

4.4.2.2 DQL Based RC Algorithm

We employ the deep Q-learning network (DQN) to find out the best policy from the action set for a given status s. Via a reward function $w(s, r)$, the long term reward from $\mathbf{Q}(s, r; \theta)$ of the DQN can be trained to converge, as shown in Fig. 4.20. The DQN weights θ can be updated by minimizing the loss function, i.e., the error of mean square between the target output of DQN, i.e., $w(s, r) + \eta \max_{r'} \hat{\mathbf{Q}}(s', r'; \theta)$, where η is the discount factor, and the present prediction $\mathbf{Q}(s, r; \theta)$. The mapping between the network context and the DQN elements, i.e., the action set that the terminal can select, environment stat s and reward function $w(s, r)$ are defined as follows.

Action Set The action set of the DRL agent includes all the rate that the transceiver can choose from, i.e., the action set is R.

Environment State The DRL agent selects the rate based on its memory according to the historical SNR values. Since the optimal selection is determined by the SNR value in the transmission slot, the environment state is defined as a set of recent SNR values in previous time. For the ith transmission, the state is defined as

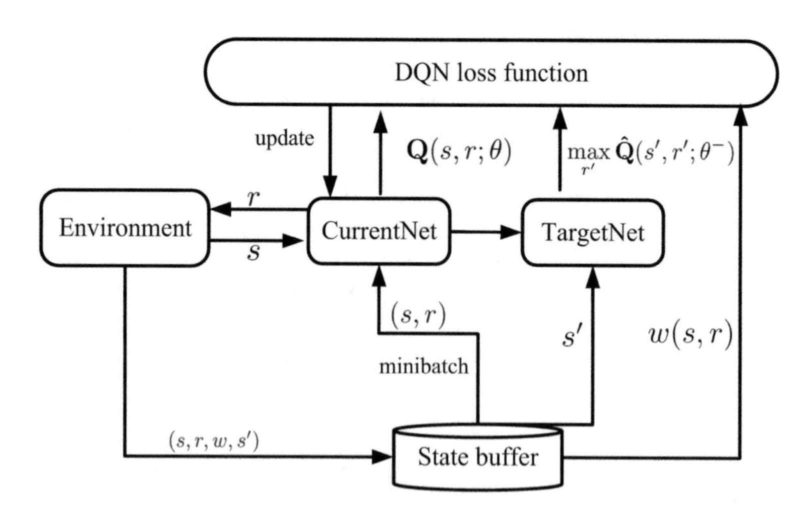

Fig. 4.20 DQN framework

$$s_i = \{SNR_{t_i-M}, SNR_{t_i-M+1}, \ldots, SNR_{t_i-1}\}, \tag{4.9}$$

where t_i is the closest time index before the transmission of the ith time slot.

Reward Function Since the objective is to maximize the average channel capacity utilization for vehicles, the reward function needs to consider not only the successful transmission of the packet, but also the channel utilization of each transmission attempt, which is defined as follows.

$$w(s, r) = \begin{cases} \frac{r}{\hat{r}_i}, & \text{If successful} \\ 0, & \text{Otherwise.} \end{cases} \tag{4.10}$$

$\hat{r}_i$ is the maximum rate that can be achieved in ith time slot, i.e., the link capacity in the transmission according to the signal strength and SNR at slot i.

As shown in Algorithm 2, for first M packets, the DRL agent randomly selects the rate for next packets. The transceiver adopts the ε-greedy policy, which selects the rate r output by the DQN, i.e., $\max_r \hat{\mathbf{Q}}(s, r; \theta)$, with probability of $1 - \varepsilon$, and randomly select a rate r from rate profile R. The agent then observes the next state and stores the experience that includes a state transit after a transmission attempt to local buffer. Then the agent randomly selects B experiences and updates the DQN weights based on the average difference values between DQN and the target DQN. And the target DQN is updated after X packets.

Algorithm 2 DRL-RC

1: Initialize two DQN with equal random weights $\theta^- = \theta$
2: Vehicle transceiver randomly select rate for M packets, records the experience (s, r, w, s') in memory buffer
3: Loop for E episode:
4: For ith time slot, select rate r_i using ε-greedy
5: Calculate immediate reward using $w(s_i, r_i)$
6: Switch to s_{i+1}
7: Store the experience (s_i, r, w, s_{i+1}) to buffer
8: Randomly select mini-batch B in buffer
9: Updates DQN weights θ by minimizing the loss
10: Updates target DQN weights $\theta^- = \theta$ every X packets

4.4.3 Performance Evaluation

In this section, we provide extensive simulations, and by comparing with traditional IEEE 802.11 RC algorithms, the link capacity utilization is shown to be significantly improved.

Table 4.6 Mobility types of vehicles

Motion type	Map size	Velocity (km/h)
Standard square	$1 \times 1, 2 \times 2, 3 \times 3, 4 \times 4$ blocks	[30, 40]
Region map	2000 m $\times$ 2000 m	[20, 30]
UAV circle	$r = 60$ m, $\ell = 20$ m, $h = 30$ m	30

4.4.3.1 Experiment Setup

We consider three types of mobility patterns, i.e., mobility of vehicles in standard square blocks streets[9] and in a region from real map data, a UAV that cruises in a circle above the ground access station, as shown in Fig. 4.19. Specifically, we consider multiple sizes of standard blocks, one real region and one UAV motion, as listed in Table 4.6.

A block size is set to be square with side length of 80 m, and the street width is set to 24 m. For standard block mobility, the vehicle travels at a speed value randomly from [20, 30] m/s after each intersection, where the vehicle will randomly turn right, left or keep the original direction. The vehicle will also stay at the intersection for a time duration uniformly distributed from [0, 10] seconds considering the traffic lights. For each vehicle scenario, a trace of 1000 s is collected and 200 s trace for UAV.

We also consider the mobility of vehicles in real region map trajectory. Two hundred vehicles' trajectories that lasts 1000 s are generated[10] based on non-standard block roads from the real map data of University of Waterloo campus, as shown in Fig. 4.21. The small circle indicates the start point of the vehicles.

The aggregated signal strength is composed of the path loss, shadowing and multi-path fading components defined in Sect. 4.4.1.1. For the access station, we employ the IEEE 802.11af radio technology that works on TV white space (TVWS) spectrum that can propagate long range (around 10 km) and penetrate through ground buildings. The carrier frequency for both vehicle and UAV terminals are set to 650 MHz. For ground vehicle terminal in standard square streets, the access station is placed on the corner that is nearest to the center of the region. For non-standard square region in Fig. 4.21a, the access station is placed at the place of (2000, 1200) in Fig. 4.21b. The ground station for UAV is placed according to the parameters in Table 4.7 and Fig. 4.19. The transmit power for both access station and the terminal is set to 30 dbm. The channel parameters are listed in Table 4.7.

The signal strength is sampled while the vehicle is moving from the received beacon frames, whose broadcast period is 10 ms. We use recent 30 historical SNR records as the state for the next frame's rate selection, i.e., $M = 30$. The rate profile of the IEEE 802.11af is listed in Table 4.8 [23].

[9] For example, the city of Portland, Houston in US, etc.

[10] Generated by VISIM software.

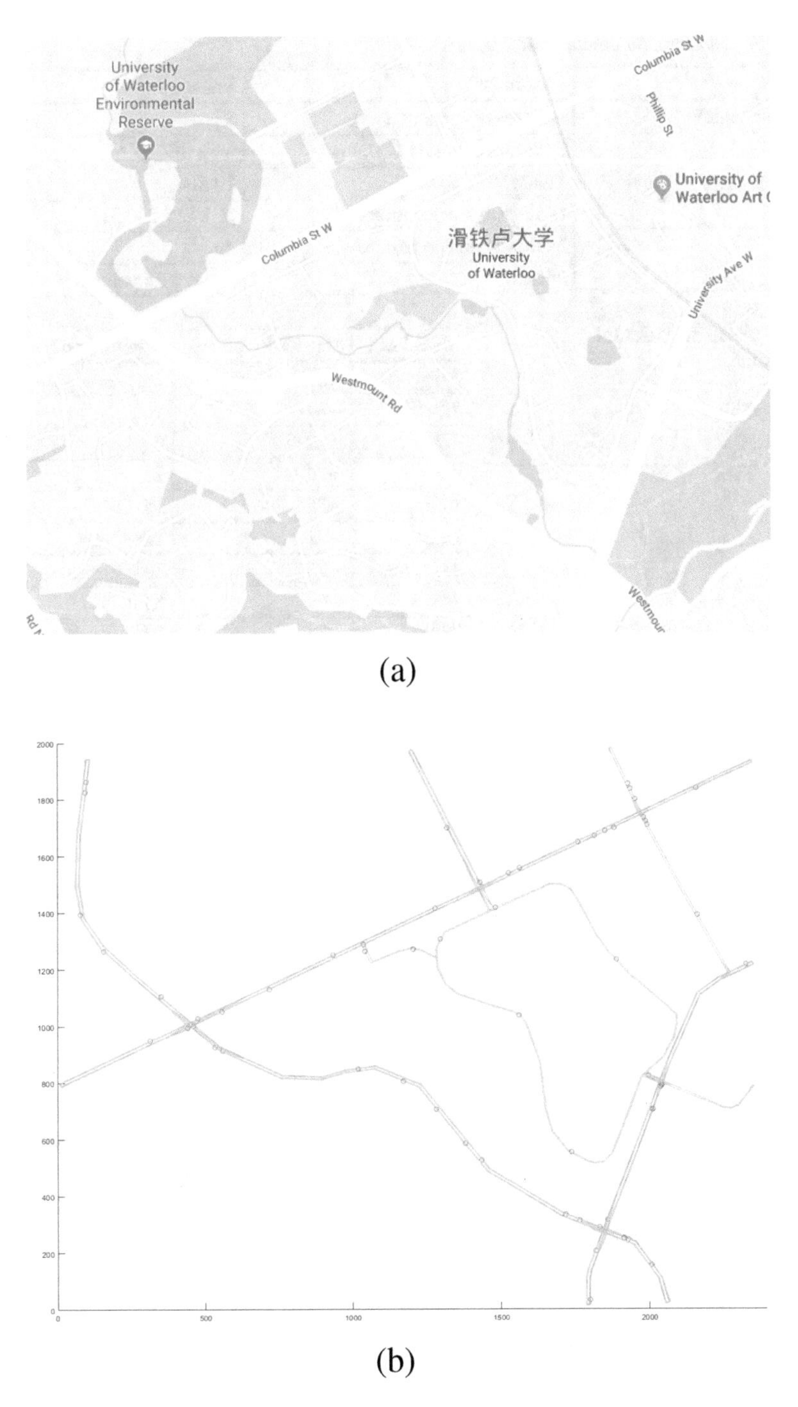

Fig. 4.21 Mobility trace for vehicle terminals in non-standard region. (**a**) UW region map. (**b**) Trajectory

Table 4.7 Network parameters for vehicles

Vehicle	Parameter	Value
Ground vehicle	Path loss exponent n	3
	Path loss constant C_{PL}	−12 dB
	Shadowing power	10 dB
	Multipath shape param.	[5 10 15 20 25]
UAV	Multipath shape param.	20
	Others	Same with vehicle

Table 4.8 IEEE 802.11af rate profile

MCS level	Sensitivity (dBm)	Minimum SNR (dB)
2.4	−87	2
4.8	−84	5
7.2	−82	9
9.6	−79	11
14.4	−75	15
19.2	−71	18
21.6	−70	20
24	−69	25
28.8	−64	29

The miniBatch size B is set to 48, and the state buffer size X is set to 100. Episode account E is set to 1500. The neural network configuration is shown in Fig. 4.22. Thirty nodes are employed for 30 SNR values input and two fully-connected hidden layers are adopted. Ten output nodes are used to give the corresponding probability of rate selection.[11] The neural weights are trained for each scenario respectively by diving the traces, i.e., half of them are used for training and the remaining half are used for test.

4.4.3.2 Performance

The average channel utilization is calculated according to Eq. (4.8) for the proposed DRL-RC algorithm. We also compare the results with the auto rate fallback (ARF) algorithm and fixed rate selection scheme (rate is fixed to 14.4 for every packet) [24]. As shown in Fig. 4.23, the proposed DRL-RC can utilize near 90% channel capacity in standard square block streets scenarios, and near 80% for both non-standard map region (UW campus) and UAV terminals.

It can be observed that the DRL-RC can improved the performance by 94% comparing with the tradition ARF algorithm, which would easily decrease the MCS level for temporally packet drop event. It can also be observed that the Fixed rate

[11] In total 9 MCS levels and plus a rate selection of 0, which indicates the transceiver should not send packet if SNR is too low.

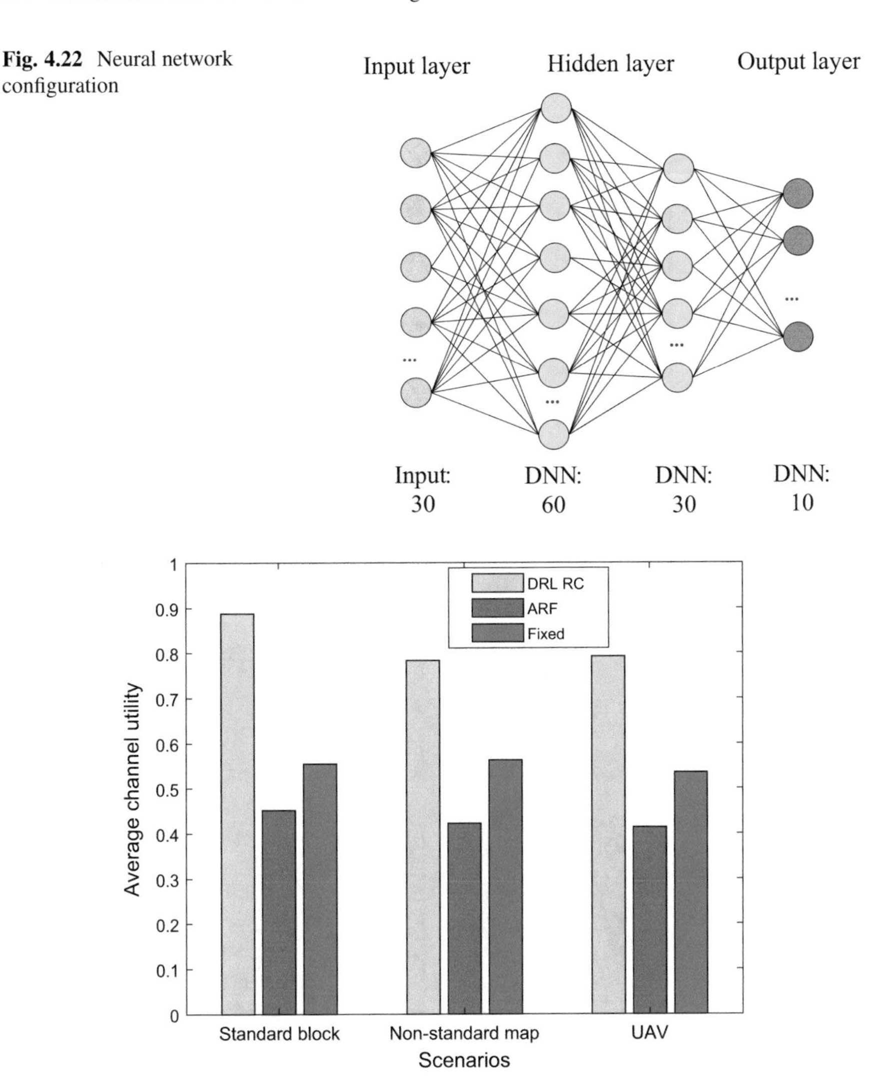

Fig. 4.22 Neural network configuration

Fig. 4.23 Average channel utility comparison

selection is also better than ARF, since it keeps selecting the same rate and thus cannot be interrupted by the transmission results.

We also compared the average channel utility for different multi-path fading shape factor K, which indicates the power ration between the LOS and NLOS components that determined by the environments. It is shown in Fig. 4.24 for the vehicle terminals in both standard square blocks streets and UW campus, the average channel utility increase slightly when K increase, i.e., when the proportion of the LOS power component increase. And for all different K values, the proposed

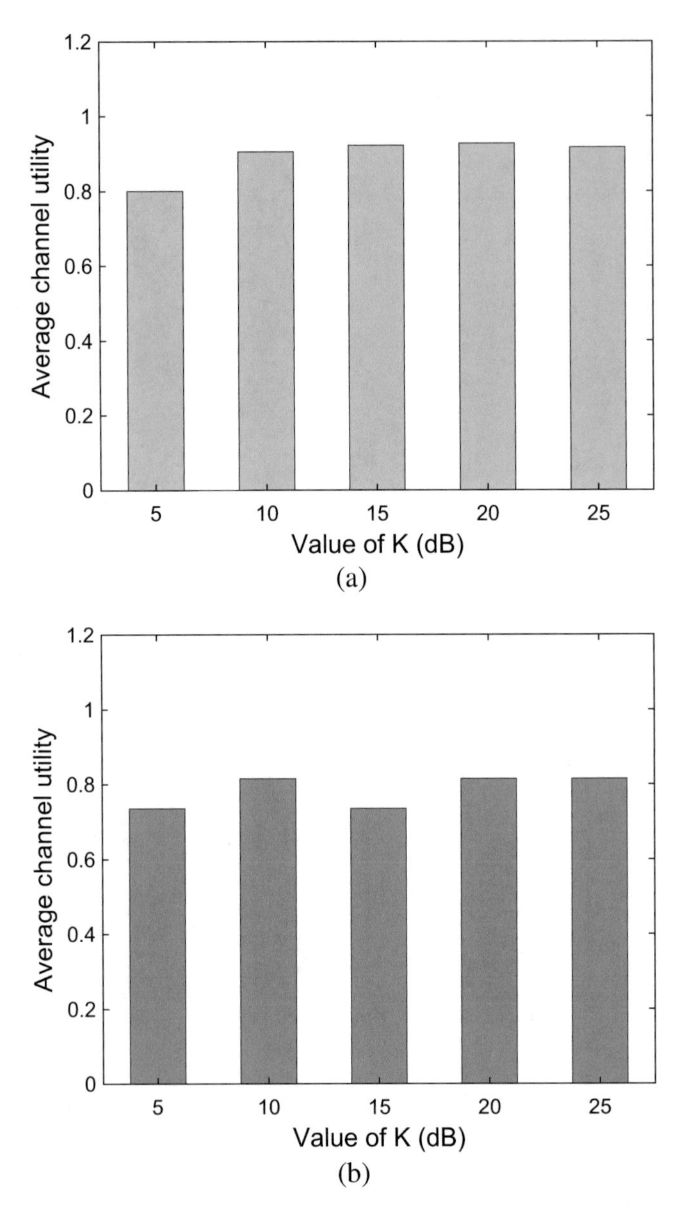

Fig. 4.24 Average channel utility vs. multi-path shape parameter K. (**a**) Channel utility for standard square blocks streets. (**b**) Channel utility for UW campus region

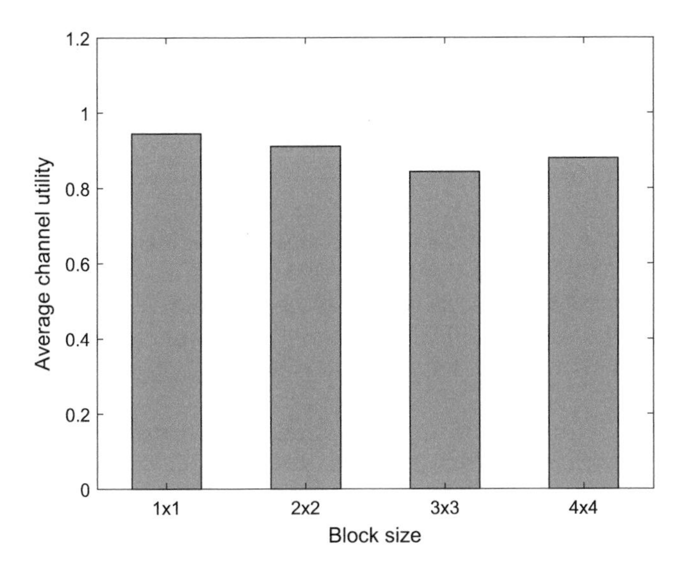

Fig. 4.25 Average channel utility vs. region size

DRL-RC can achieve significant performance, which shows that it is robust to different multi-path fading levels.

We also compared the DRL-RC performance for different size of the regions that a vehicle terminal travels. As shown in Fig. 4.25, setting the shape parameter to 20 and it can be observed that when the region size increases from 1×1 blocks to 4×4 blocks, the average channel utility fluctuates slightly around 90% of the link capacity. Such result show that the DRL-RC is scalable to the mobility range, and thus the scope of the DRL-RC application can be extended.

4.4.4 Summary

In this section, we have investigated the automatic rate control for vehicles with more general mobility using the DRL method. The signal variations in different motion styles are exploited both in simulated mobility trace and real map data. It is shown that the channel utility can be significantly improved for different channel conditions and mobility area size. The proposed method can be extended to more general case of Internet of mobile things.

4.5 Intelligent Rate Control for Internet of Maritime Vehicles

Marine resources are important to human society, and the Internet of Maritime Vehicles can provision the sustainable development in maritime conditions [25]. In this section, we consider to provide the TVWS access for maritime vehicles using the vacant TV bands. Campos et al. prove that the 700 MHz TVWS band can provide both cost-effective and high data rate communication between ships and shore infrastructures [26]. TVWS can provide around 100 km wireless access, and thus are very suitable to envision Internet access for maritime vehicles [27, 28]. Compared with the long-rage WiFi, the IEEE 802.22 based TVWS communication is proved to be more feasible for maritime communication due to its long propagation capability [29].

There is an critical issue to apply the IEEE 802.22 based TVWS spectrum in maritime communication, which is to select proper modulation and coding scheme (MCS) and corresponding rate for every packet to send. Selecting too large rate will cause high packet drop rate due to the channel capacity limitation, while selecting too small rate will compromise the spectrum efficiency. Unlike cellular networks, IEEE 802.22 has not specified a dedicated control function and channel to monitor the data link status continuously[12] and leave the MCS selection to users' discretion. Conventional MCS selection schemes for distributed networks are often based on the statistics of transmission history for stationary or quasi-stationary users, e.g., the auto rate fall back (ARF) scheme, Minstrel, etc. which are not applicable in mobile conditions, especially for maritime vehicles in complex maritime environment [8]. Efficient MCS selection should adapt to the channel variations, which is determined by three main factors. The dominant part to cause signal attenuation is the path loss that determined by the transmission range between the MIoT terminal and the TVWS access station, which keeps changing due to the mobility of the terminal. Besides, the wireless signal will vary with channel shadowing caused by sea wave, other ships, etc. In addition, the multi-path interference from multiple reflections in maritime environment can also impair the received signal strength. Due to the complexity in maritime environment, it is difficult to setup explicit model to accurately evaluate the channel condition to select proper MCS. Existing channel measured references, such as received signal strength indicator (RSSI) and channel state information (CSI), are from short-term signal measurement and cannot directly predict the channel capacity for the imminent transmission due to short channel coherent time, especially in mobile conditions [9].

As the maritime vehicles often travel along fixed sea lanes due to safety concerns and transportation efficiency reasons, e.g., to avoid submerged reef and to fuel consumption, the transmission distance variations for a maritime vehicle to travel the same route have similar patterns, and the consequent signal attenuation also varies similarly since the path loss dominates the received signal strength oscillation.

[12] For example, the PUCCH in LTE networks.

Wang et al. investigate the maritime channel modeling and conclude that the near-sea-surface channels are highly dependent on the location [30].

Recent success of data analytics has prompted data-driven and learning-based method to predict the received signal strength from previous experience of maritime vehicle and its brethren [31]. Joo et al. propose that the historical CSI measurements can be used to train an NN to predict the vehicular channel condition effectively [32]. Sandoval et al. show that by training a neural network (NN) using reinforcement learning, the optimal transmission parameters can be selected for LoRa Internet of things terminals [33]. Xu et al. propose that the historical CSI measurements can be used to train an NN to predict the vehicular channel condition effectively [34]. Given the location information of both the maritime vehicle and land access infrastructure, it is possible to estimate the path loss based on the transmission distance, as well as highly location related shadowing effect in maritime environment.

We consider the IEEE 802.22 based TVWS access for maritime vehicle to envision large scale and high throughput wireless access. In order to better utilize the channel capacity, we model the SNR level prediction as a time series forecasting problem. The predicted SNR level is used to choose proper MCS for the next egress frames. Specifically, a nonlinear autoregressive exogenous neural network (NARXNN) is used to relate the historical status of a maritime vehicles and the SNR level in the imminent time slot. The historical status includes both the records of recent transmission distances and SNR levels. It is shown that the results of this forecaster are useful for selecting the optimal MCS level and improve the channel resource utility.

- Ab intelligent MCS selection algorithm is proposed for the wireless link between an maritime vehicle and a land access station. Specifically, we consider an uncoded communication system that works in TVWS spectrum and adopts the basic MCS profile from IEEE 802.22 standard, and design a location aware and data-driven forecaster to predict the future SNR level for MCS selection of egress packets. Instead of applying explicit mathematical model that requires the a-priori knowledge of the system, we apply an NARXNN based black-box modeling that utilize a certain size of discrete historical samplings of transmission distances as well as the channel SNR levels to find out the channel status in the next transmission moment, which is used to select the MCS level accordingly.
- Efficient training architecture is designed by combining several distance and SNR samplings set from several trips along the same sea lane to form a trace set with periodicity that is suitable for NN to find out the long-term SNR variations. During the training phase, we apply the open loop (series-parallel) architecture that both the distance and SNR are practical, while in prediction phase, the close loop (parallel) architecture is applied to compensate incomplete SNR sampling history.
- Extensive simulations are conducted and the results show that the average channel utility can reach 85% of which from the optimal MCS selection scheme,

and is up to 70% higher than traditional sampling based scheme. By comparing the performance on different NN size and Lag order, we also draw useful conclusions about the relationship between the network performance and the NN complexity, which is useful for efficient channel forecasting and select optimal MCS for mobile IoT terminals.

4.5.1 Related Works

4.5.1.1 TVWS Access for MIoT

TVWS refers to the inactive or unused spectrum spans from 460 to 700 MHz, which can provide broadband and long-range wireless access without harming primary television users. The application of TVWS in maritime communication has been investigated in [26–28, 35]. Compared with conventional terrestrial communication methods, such as cellular and WiFi networks, TVWS can provide better coverage and Non-Line-of-Sight (NLOS) Performance for its low carrier frequency. And compared with satellite communication, TVWS can provide low-latency wireless access with much less cost by employing an offshore land access station. Such advantages of TVWS access for maritime vehicle in offshore regions have been investigated in [29, 36]. IEEE 802.22 standard specifies the physical (PHY) and medium access control (MAC) layers to envision long range wireless region area networks (WRAN). However, the standard has not specified a rate adaptation scheme to dynamically choose the MCS level according to the channel variation, which is crucial for secondary users in TVWS, i.e., to utilize the channel resource when primary users are absent or inactive [37, 38].

4.5.1.2 MCS Selection Schemes

For network systems that have no control plane to monitor the channel status, the transceiver has to autonomously adjust the MCS level according to other channel references. There are two categories of such MCS adjust algorithms, i.e., statistic-based and measurement-based types.

Statistic-Based MCS Selection The transmission results are recorded and some reference metrics, e.g., packet error rate (PER) for a certain recent period or for specific MCS level, are calculated to evaluate the channel conditions and predict the achievable MCS. Such mechanism is shown to be inefficient in mobile conditions. Joshi et al. perform ground test and show that statistic-based MCS selection scheme can cause much transmission attempts due to unwise rate attempts for mobile users on road [39]. It is shown that only half of the bandwidth resource can be utilized from roadside WiFi hotspot for vehicles due to the high packet loss caused by the inefficiency of existing statistic-based RA schemes [40]. It is also reported in [41] that for mobile conditions, conventional statistic MCS selection schemes including SampleRate, AMRR, Minstrel, and RRAA, etc., have poor performance in vehicular conditions [42].

Measurement-Based MCS Selection The RSSI or CSI values are measured from transceiver sensor to infer the channel status, which is used to estimate the PER for different MCS selection. For example, the RSSI read from the RTS/CTS frames are used to estimate the channel quality and corresponding optima rate, and provides better performance than measurement-based ARF scheme [43]. To avoid the non-trivial bother to modify the RTS/CTS protocol, Judd et al. evaluate the channel by utilizing the channel reciprocity to response to the variation to select proper MCS [44]. However, since the RSSI is based on the average signal strength of multiple sub-bands, and thus can lead to inaccurate channel reference [9]. In contrast, the CSI is used to provide more accurate channel information [16], however, only a few network transceivers can provide CSI information, which limits its usage especially in TVWS access for maritime vehicles [45].

Both of these traditional statistic and measurement-based MCS selection schemes cannot directly apply in mobile condition, as the channel prediction requires non-negligible time duration when the terminal may travel to new location that the channel condition varies drastically [46, 47].

4.5.1.3 NARXNN Forecaster

The NARXNN is widely used to forecast the future status of a system which has non-linear variations. Fleifel et al. utilize a NARXNN forecaster to predict the primary user activity in LTE cognitive radio networks and show that the resource block occupancy status can be better predicted than conventional methods [48]. Narawade et al. use the NARXNN to predict the sensor network congestion and adjust the link rate accordingly [49]. In [50] the authors apply a NARXNN to predict the direct solar radiation according to several training phases, which provide better prediction if the NN training is performed periodically. Since the maritime vehicles are often moving along similar sea lanes, if the historical records can be stored, it is possible to predict the future channel variations by merging past location and the SNR information via a NARXNN.

4.5.2 System Model

We focus on an MIoT communication scenario as shown in Fig. 4.26. We consider one maritime vehicle in offshore region, e.g., boats, ships, etc. that travel within rage of 100 km from the coastline. The maritime vehicle travels along certain sea lanes and keeps on sending fixed-length packets to the land TVWS access station using the uncoded MCS profile from a IEEE 802.22 transceiver. When the terminal

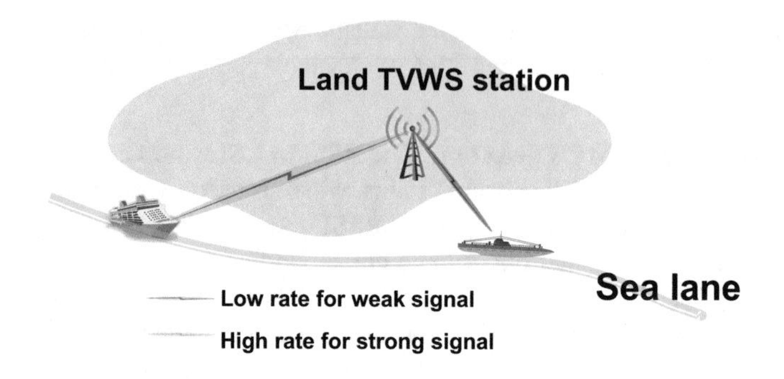

Fig. 4.26 System model for LA of MIoT

is approaching to the station, it should increase the MCS level to improve the channel efficiency, while when it leaves away from coast, lower MCS level should be selected to reduce PER.

4.5.2.1 Network Model

We assume that secondary MIoT user can periodically access to the TVWS channel when it is not occupied by primary users. Without loss of generality, we assume that the maritime vehicle is allowed to transmit once per T units of time.

The SNR level that characterizes the channel condition is determined by the transmission power, channel fading and white noise interference. We assume both the power of the transmission signal and the white noise interference are stationary.

The channel fading of the wireless link in maritime environment includes three parts, i.e., the path loss, shadowing and multi-path fading, which determines the SNR of the transmission link. In maritime environment, traditional logarithmic path loss pattern cannot fit the local oscillations of the wireless channel. Thus, we adopt two widely considered path loss models, i.e., the 2-ray and 3-ray models that can capture the deep nulls (signal sudden drop) due to the reflected ray from the sea surface (2-ray) and extra refraction due to the evaporation duct (3-ray) [30]. Specifically, the path loss of the 2-ray model can be expressed by

$$PL_{2r}(h_r, h_t, d) = -10log_{10}\left((\frac{\lambda}{4\pi d})^2(2sin(\frac{2\pi h_t h_r}{\lambda d}))\right), \tag{4.11}$$

where h_t and h_r are the height of the transmitter and receiver antennas, respectively, λ is the wave length of the signal carrier, d is the distance between the maritime vehicle and the access station.

According to the analysis in [51], the probability that the third path exists is around 8.5%, thus the 3-ray path loss model can be obtained from

$$PL_{3r}(h_r, h_t, h_e, d) = -10log_{10}\left((\frac{\lambda}{4\pi d})^2(2(1+\Theta)^2) \right), \tag{4.12}$$

where h_e is the height of the effective evaporation duct, and Θ can be obtained by

$$\Theta = 2sin(\frac{2\pi h_t h_r}{\lambda d})sin(\frac{2\pi(h_e - h_t)(h_e - h_r)}{\lambda d}). \tag{4.13}$$

The signal variations with distance for both 2-ray and 3-ray models are shown in Fig. 4.27a, where deep nulls can be observed for certain distance range between the MIoT and the access station. Although shadowing effect is not obvious due to the sparse obstacles in maritime conditions, it is observed that such slow-fading follows the log-normal distribution, which is applied in our network model considering that many maritime vehicles coexists in offshore regions, as well as different sea states [21, 52].

Apart from the dominant transmission paths in 2-ray and 3-ray model, we also consider multi-path interference from other reflections and scatterings, which is assumed to follow the Rice distribution. The shape parameter K is defined as the ratio of power contributed from dominant paths to the remaining components of the received signal. To show the impacts of different multi-path fading levels to the performance of the proposed MCS selection scheme, we consider different values of K that ranges from 15 to 25, as the multi-path interference is also sparse in maritime conditions.

All these three channel fading effects are superimposed to impair the transmission signal and alter the corresponding SNR level, which lead to different PER for different MCS selections.

4.5.2.2 MCS Profile

The transmission result is determined by both the uplink SNR level during the transmission time slot and the packet's MCS selection. We consider four different uncoded MCS levels according to IEEE 802.22 standard,[13] as shown in Table 4.9, where ρ is the symbol SNR value [53]. The number of symbols for a packet with length L can be calculated by

$$N_{symbol} = \lceil \frac{L}{m_i} \rceil, i \in [1, 4], \tag{4.14}$$

where m_i is the number of bits within a symbol of the corresponding ith MCS level. And the packet error rate (PER) is determined by

[13] The standard includes multiple types of coding methods, we consider uncoded system and the link rates are calculated accordingly.

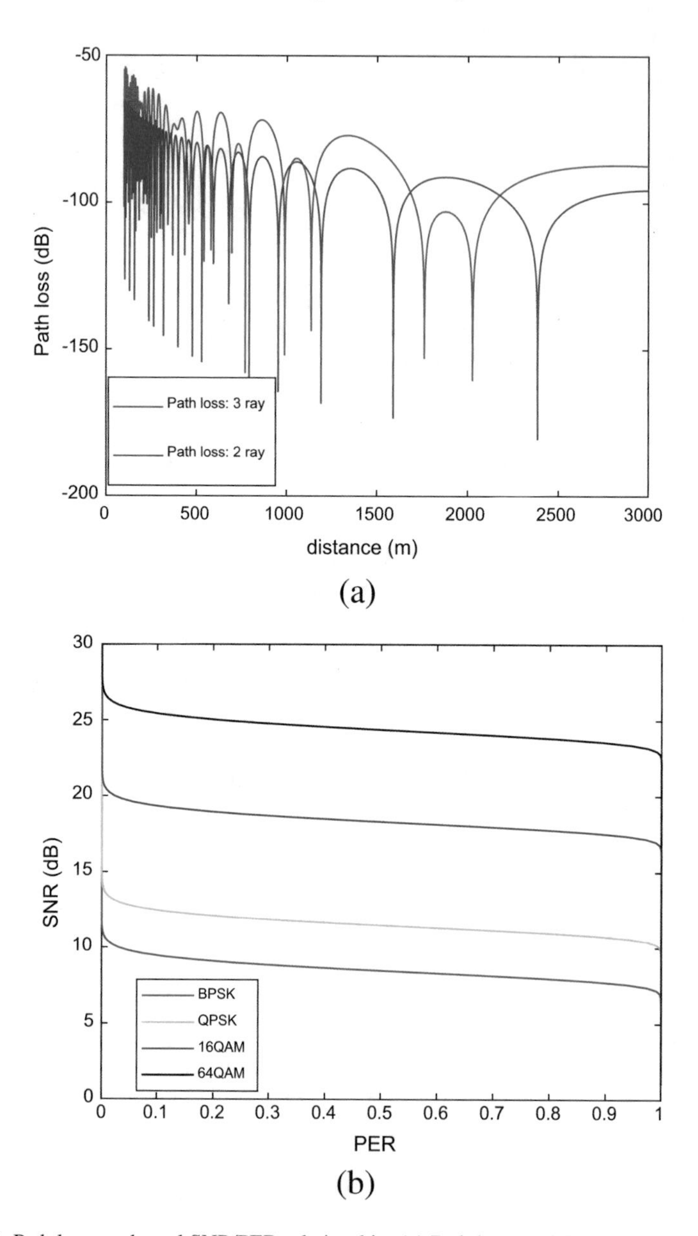

Fig. 4.27 Path loss mode and SNR/PER relationship. (**a**) Path loss models: 2-ray 2 vs. 3-ray. (**b**) Required SNR vs. PER

$$PER = 1 - (1 - SER(\rho))^{N_{symbol}}, \qquad (4.15)$$

where $SER(\rho)$ is the symbol drop rate obtained from Table 4.9 given the symbol SNR ρ. The relationship between PER and the SNR is plotted in Fig. 4.27b.

Table 4.9 MCS profile of the uncoded system based on IEEE 802.22

MCS level	Modulation	Uncoded data rate R	SER [54]
1	BPSK	6 Mbps	$Q(\sqrt{2\rho})$
2	QPSK	9 Mbps	$2Q(\sqrt{\rho})$
3	16QAM	18.16 Mbps	$3Q(\sqrt{\frac{3}{15}\rho}) * (1 - \frac{3}{4}Q(\sqrt{\frac{3}{15}\rho}))$
4	64QAM	27.22 Mbps	$3.5Q(\sqrt{\frac{3}{63}\rho}) * (1 - \frac{7}{8}Q(\sqrt{\frac{3}{63}\rho}))$

4.5.3 Proactive NARXNN Forecaster Based MCS Selection

The MCS selection strategy can be made from effective channel condition forecasting. It is known that the path loss takes up the majority of the channel attenuation, i.e., the components that changes with the distance variations when the maritime vehicle travels along the sea lane. Since Terminals travel in the same route will have similar variation patterns, which can be recognized by training a neural network to fit the function on historical channel observations and the transmission range given by the location of the maritime vehicle. Assume the channel condition is symmetrical and the maritime vehicle can periodically obtain the SNR information, e.g., from the received beaconing frames broadcast by the access station. A series of the historical SNR values are maintained by the terminal as a vector S of length D (D is also called the lag order). We consider D downlink SNR values sampled at discrete time t, which is denoted by $S(t)$, while D corresponding distance values are also recorded and denoted by $P(t)$. Then the MCS selection can be defined as a function f that maps from a series of historical SNR and transmission distance values to the channel SNR of the next time moment:

$$\hat{S}(t) = f\Big(P(t-D), P(t-D+1)\ldots P(t-1), \tag{4.16a}$$

$$S(t-D), S(t-D+1)\ldots S(t-1)\Big), \tag{4.16b}$$

Since the mapping function f is determined by both the channel fading and the MIoT mobility, it is difficult to setup explicit model or close formula for f. The most standard way for time series prediction is to use the recurrent neural networks (RNNs), which have been extensively applied in the machine learning community for time sequence modeling, like speech recognition [55], the natural language modeling [56] and traffic prediction [57] etc. When a vanilla RNN model is employed, the SNR value at the time step t can be predicted by the observed SNR value, distance and the RNN's hidden state h_{t-1} from the previous time step $t-1$. The prediction process can be mathematically expressed as

$$h_t = tanh(W_{hh}h(t-1) + W_{xh}x_{t-1}), \tag{4.17}$$

$$\hat{S}_t = W_{hy}h_t, \tag{4.18}$$

where $x_t \triangleq [P(t-1), S(t-1)]$ is the vector consisting of the observed distance and SNR value at time $t-1$; W_{hh}, W_{xh} and W_{hy} are the model parameters to be learned from the training data; $tanh(\cdot)$ is the hyperbolic nonlinear activation function. The model parameters, along with the initial hidden state h_0 can be initialized randomly. Obviously, the future SNRs can be predicted step by step using the RNN model. However, vanilla RNNs are well known by their limited ability to capture long-distance correlations. In the marine application scenarios, the channel status often exhibits long dependencies, that is, the channel SNR at time t is often correlated with the SNRs at tens of time steps before, which makes the vanilla RNNs not suitable to be used here. Some variants of the vanilla RNN have been proposed to capture longer dependencies, such as the long short-term memory (LSTM), but these models are often very complex and thus are very easy to be overfitting when the training dataset is not large enough. In order to being able to capture the long dependencies while avoiding the use of too complex models, we instead propose to train an NARXNN to approximate the function f, and predict proper MCS level that should be selected for next egress frame. As shown in Fig. 4.28, during the training phase, both of the location information and SNR observations are recorded and the open-loop architecture is applied as the true past inputs are available[14] and more precise model can be achieved. And in the testing phase, the close-loop is applied for multi-step-ahead forecast, and also the past channel samplings might not be complete. Thus the training process is to adjust the NN coefficients to minimize the difference between $\hat{S}(t)$ and $S(t)$. The historical SNR and corresponding transmission distance when the maritime vehicle travels from one to another side of a certain sea lane are combined together, which is defined as single-trip trace set, and $S(t)$ and $P(t)$ includes several single-trip trace sets, e.g., if the maritime vehicle travels along the sea lane for X times and all the historical records are combined together and called X-trips trace set.

To utilize the cognitive TVWS spectrum resource, the airtime to transmit a packet should be as small as possible to improve the channel utility and reduce potential interference to primary users. We focus on the average channel utility (ACU) of the proposed LA, which is crucial when utilizing the cognitive TVWS resources, and it is defined as

$$U = \frac{1}{Y}\sum_{i=1}^{Y}\frac{R_i}{R_{opt}} * z_i, \tag{4.19}$$

where Y is the number of total transmission attempts, R_i is the MCS selected for the ith transmission attempt based on the prediction $\hat{S}(t)$ using Eq. (4.16), R_{opt} is the optimal MCS rate according to the channel status.[15] And z_i is the transmission

[14] Relying on survey vessel or select complete traces from data collection mechanism such as crowd sourcing.

[15] Optimal values are obtained as the simulated channel condition can be obtained, while it is unknown to the LA algorithm.

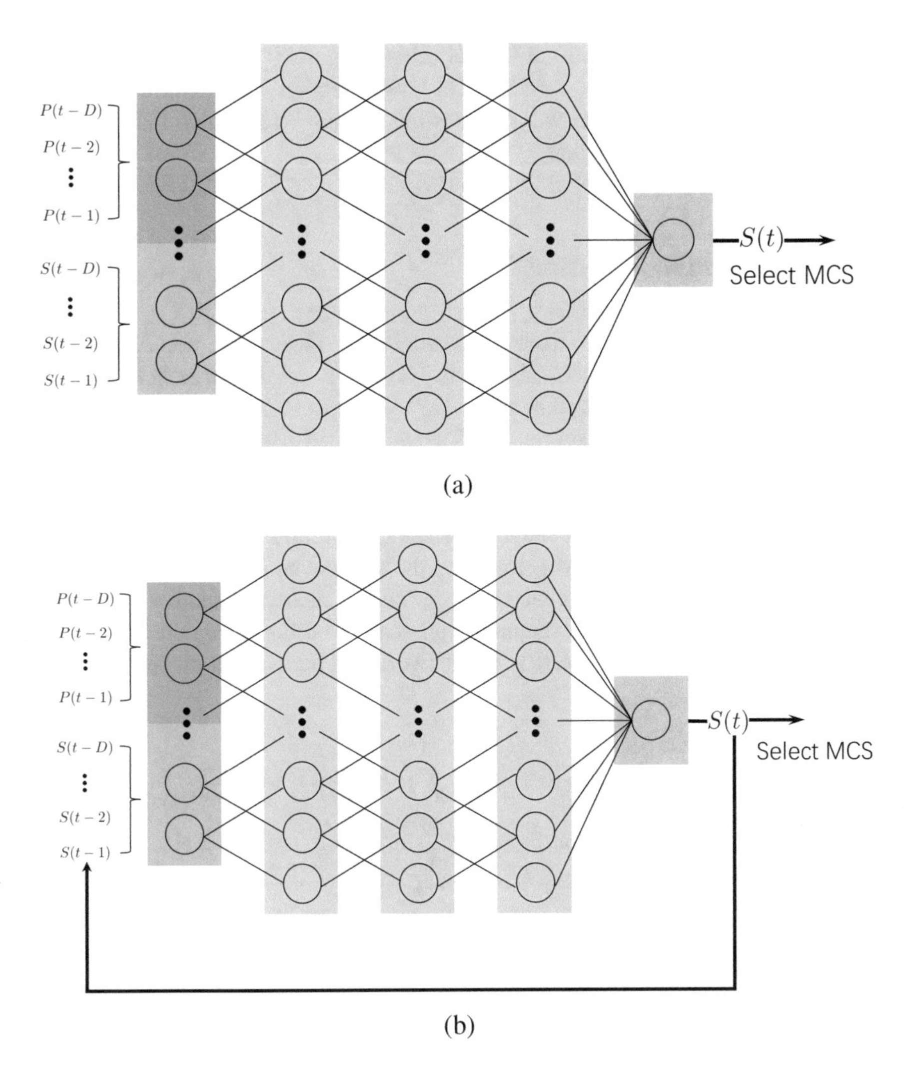

Fig. 4.28 Nonlinear autoregressive neural network architecture. (**a**) Open loop architecture for training. (**b**) Close loop architecture for testing

result obtained from Eq. (4.15) and equals to $1 - PER$. Maximizing U means to select the rates that are closest to optimal ones, thus to minimize the transmission airtime. The received SNR series together with the corresponding distance series are recorded and collected from the terminal or its brethren. Such SNR series set is used to train the NN in Fig. 4.28 by minimizing the selected rate R_i and the optimal rate R_{opt} for ith transmission following the Levenberg Marquardt algorithm [58]. When applying the NN for the proposed LA, the NN first input D historical SNR values, and output a predicted SNR $\hat{S}(t)$, which is then used to select the maximum MCS that keep the PER less than a negligible value. Since the PER is almost a step function of SNR shown in Fig. 4.27b, this value is set to 10^{-3}. When the predicted SNR is too low that even the lowest MCS level selection would cause significant packet loss, so the transceiver will not transmit when the calculated PER from the estimated SNR is non-negligible.

4.5.4 Performance Evaluation

We have conducted mobility-trace-drive simulation to evaluate the ACU of the NARXNN prediction based MCS selection. The MIoT mobility trace in Hong Kong offshore area is collected from the online AIS service website [59]. Specifically, We consider three sea lanes marked in Fig. 4.29 and consider the TVWS access based on IEEE 802.22 for the maritime vehicles travel from the one side to another of a certain sea lane. As many terminals travel along the same lane day by day, the SNR traces can be collected per trip, and aggregated at a central server, e.g., at the access station or other edge servers. We simulated the channel condition and collect a set of SNR and distance records of each sea lane, half of them are used for training the coefficients of the NN, and the remaining parts are used for testing the ACU of the MCS prediction given in Eq. (4.19). The network and the NN parameters are listed in Table 4.10.

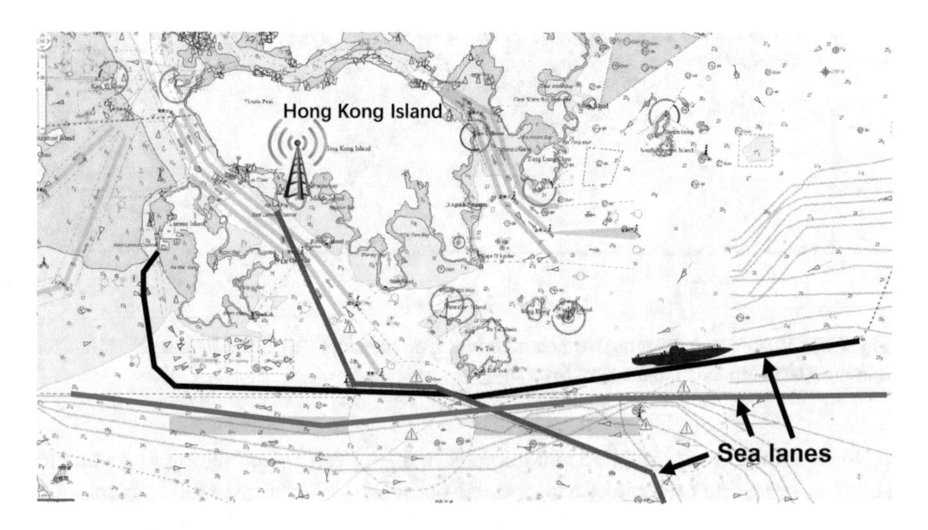

Fig. 4.29 Mobility along sea lane in Hong Kong's surrounding waters

Table 4.10 Network and NN parameters

Parameters	Value
Carrier frequency	650 MHz
Bandwidth	6 MHz
MIoT transmission power	1 w
Power of shadowing	5 dB
Power of white noise	−150 dBm/Hz
Shaping parameter K	[15, 20, 25]
h_t	100 m
h_r	11 m
h_e	12 m
Packet length L	1000 Bytes
Transmission cycle	10 ms
SNR measurement cycle	10 ms
MIoT velocity	12 nautical mile/h
Number of NN layers	[2, 7]
D	[5, 40], step size of 5

4.5.4.1 Channel SNR Prediction

Figure 4.30 shows the performance of the channel SNR prediction of the proposed NARXNN method.[16] It is observed in Fig. 4.30a the differences between $S(t)$ and $\hat{S}(t)$ are within 5dB, which is comparable to the SNR step between consecutive MCS levels as shown in Fig. 4.27b, and thus is sufficient to estimate the proper MCS level. Figures 4.30b and 4.30c show a close observation of the SNR prediction results for both $K = 10$ and $K = 20$ cases. For both the two cases, the predicted SNR can well capture the signal fluctuation. When $K = 20$, the multi-path interference is less significant, which lead to better prediction results.

4.5.4.2 Lag Order Impact

The size of the historical inputs, i.e., the lag order D, should be properly selected. Too large lag order may require significant training overhead, while too small lag order may lead to poor performance. The relationship between the ACU performance and the lag order is given in Fig. 4.31 when three hidden layers are employed. It can be observed that the ACU U grows from near 70% to 85% when the lag order D increase from 5 to 16 for both two path loss model, and then fluctuate slightly when D continue to increase to 30, which demonstrates that around 16 historical inputs are sufficient to let the NN to predict the channel SNR within

[16] Lag order is set to 16, number of the NN hidden layer is set to 3 and 15-trips trace set is used.

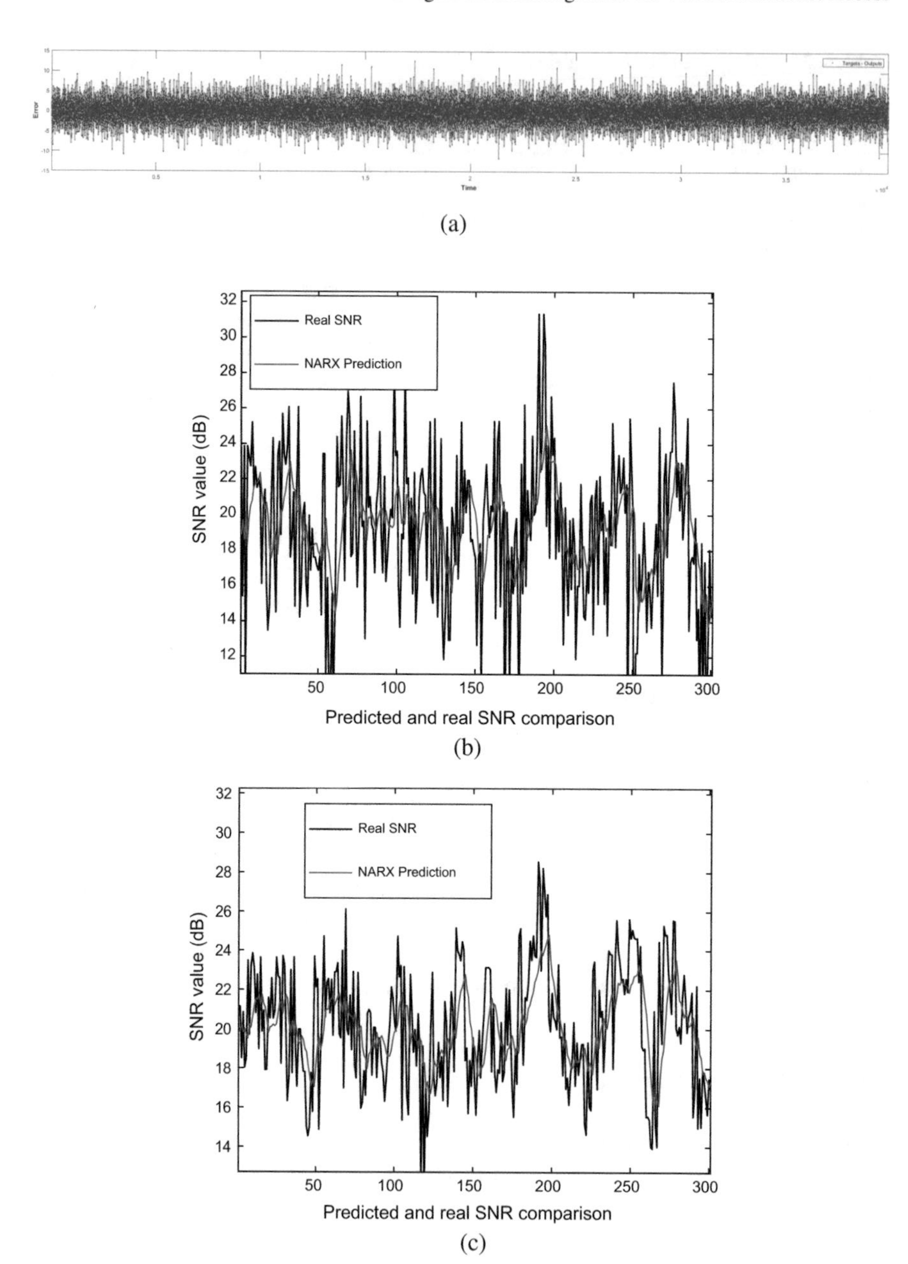

Fig. 4.30 Channel SNR prediction of the NARXNN. (**a**) SNR prediction error. (**b**) SNR prediction for K = 10. (**c**) SNR prediction for K = 20

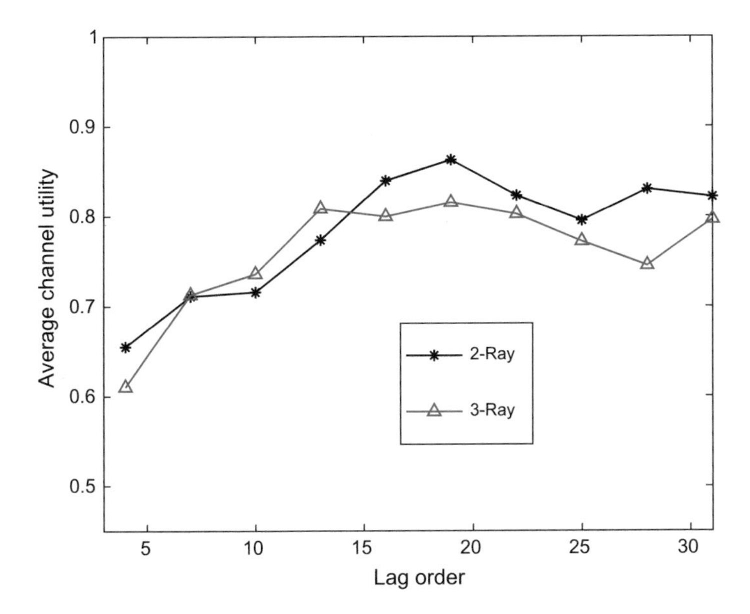

Fig. 4.31 Average channel utility vs. lag order

acceptable threshold. A larger lag order can hardly provide extra utility gain, as too old channel information provides limited additional reference than the recent SNR records.

4.5.4.3 NN Size Selection

To configure an efficient NN that is not too complex yet effective, the number of hidden layers and corresponding ACU performance are investigated as shown in Fig. 4.32. The performance of the 2-ray path loss model is compared by configuring different number of NN layers, including several hidden layer and one output layer.[17] It is shown that there is marginal performance improvement when the number of the hidden layers are more than three. Employing a three hidden layer NN already achieves the ACU around 85%. Since the marginal benefit of continually increasing the number of hidden layer is minor while leads to more computing overhead in both training and applying phase, we apply four layers NN model (one output layer and three hidden layers) for the remaining evaluation.

[17] The input layer is not counted.

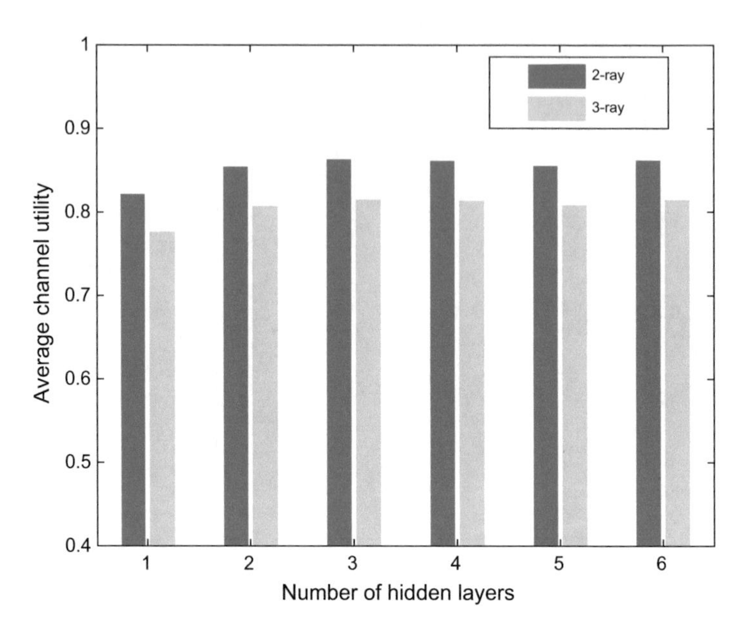

Fig. 4.32 Average channel utility vs. NN layer

4.5.4.4 ACU Performance

We apply three hidden NN layers and the lag order D is set to 15, and the average channel utilities for both 2-ray and 3-ray models of the proposed LA are shown in Fig. 4.33a and b, respectively. It is shown that 80–90% of the channel capacity can be utilized based on the NARNN prediction and corresponding MCS selection. The performance of 3-ray path loss model is slightly less than the 2-ray model, as 3-ray model is more complex and has more deep nulls.

We have also compared the performance of the proposed LA with a traditional statistic based sampling scheme[18] and a dummy policy that always choose the fixed MCS, i.e., the QPSK. It is shown that the proposed MCS selection scheme is much more efficient than the other two methods. The sampling method is too aggressive to decrease the MCS level, i.e., decrease the MCS level immediately when there is a transmission error, and too conservative to increase the link rate, which is allowed only when a set of consecutive transmissions are successful. And the next selection is limited to only the neighboring levels. For the fixed scheme, it will waste the channel capacity if the channel condition is good, while increasing the packet drop rate if channel gets worse. The NN can learn from past experience, and effectively

[18] The sampling scheme will start from the lowest MCS level, and increase to next higher lever if there are more than 10 consecutive successful transmission, and will reduce to lower lever if there is a transmission error.

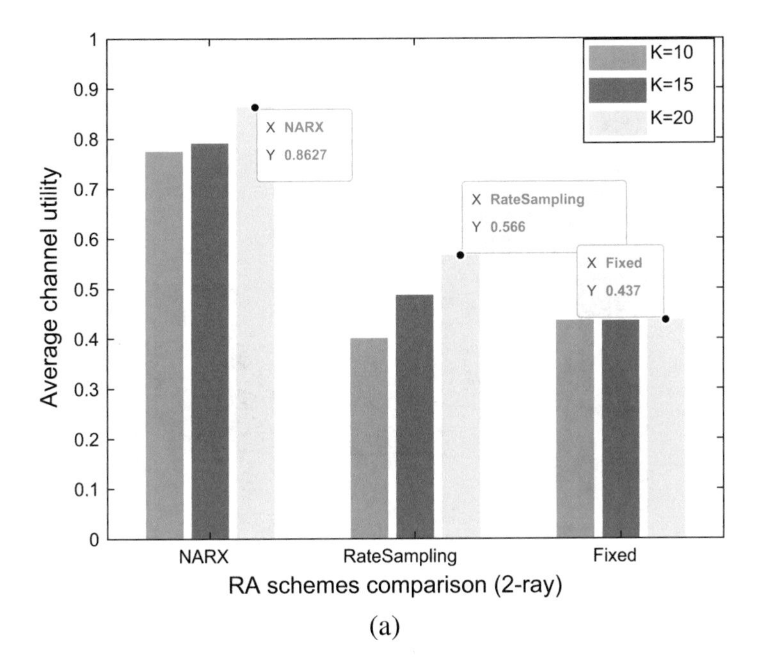

(a)

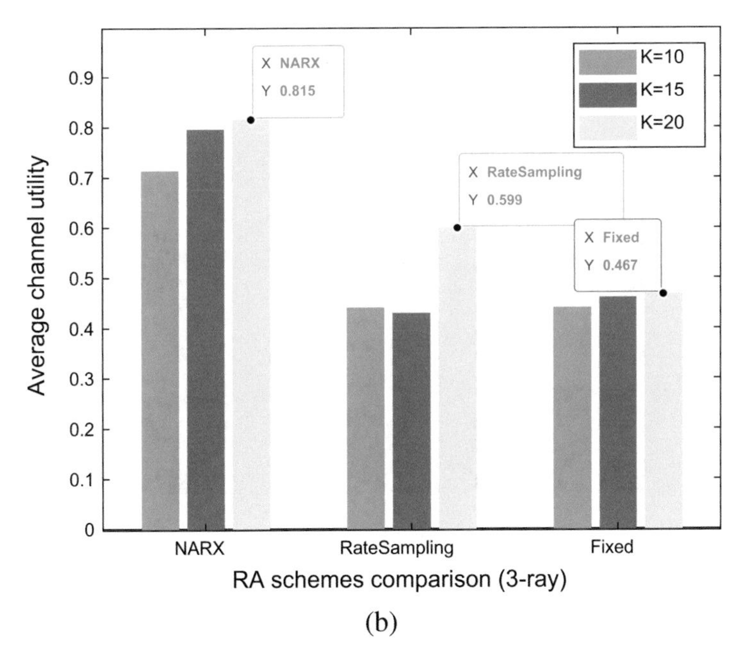

(b)

Fig. 4.33 ACU performance of the NARXNN prediction based MCS selection. (**a**) ACU performance of ray 2 path model. (**b**) ACU performance of ray 3 path model

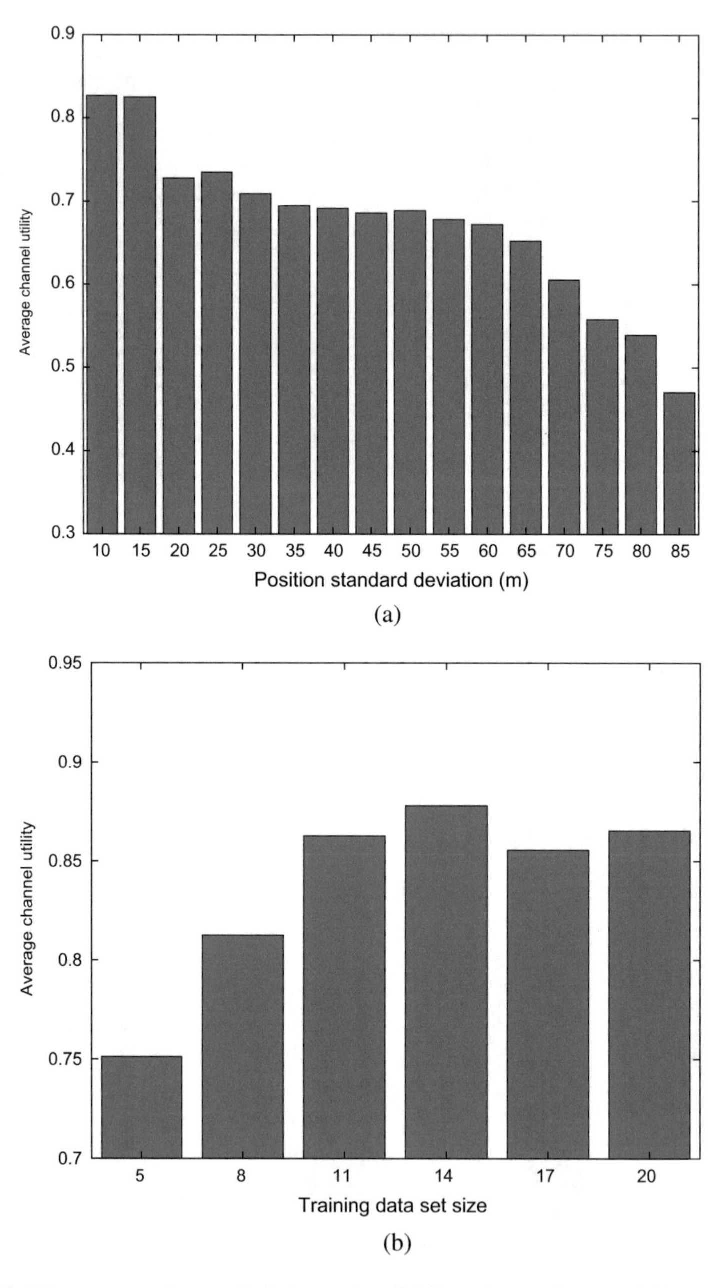

Fig. 4.34 Tolerance to trajectory deviation and availability of trace data. (**a**) ACU vs. position deviation. (**b**) ACU vs. size of the training data set

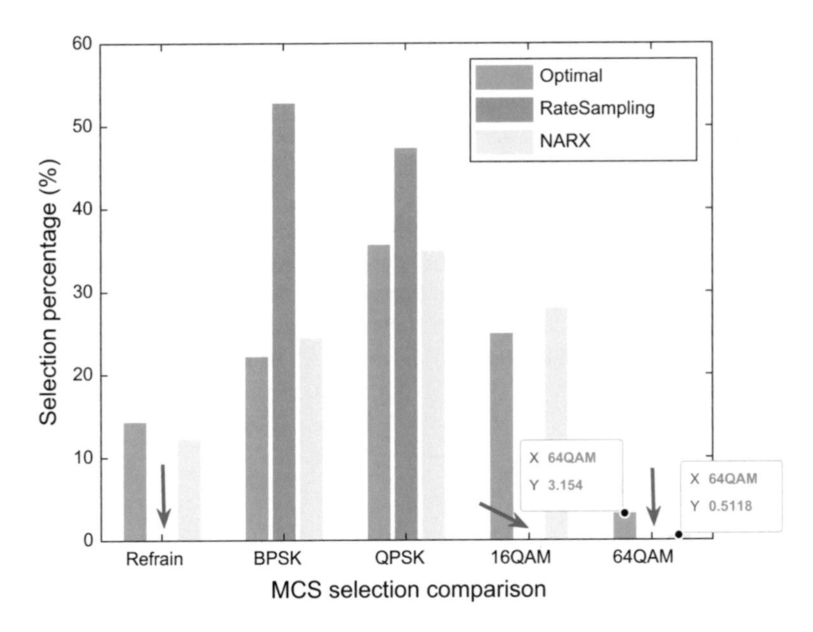

Fig. 4.35 MCS selection distribution comparison

predict the channel status within reasonable error range, and thus can precisely predict the proper MCS that should be selected for the next egress frame (Fig. 4.34).

The channel utility performance is also evaluated for different multi-path fading levels. We set different K values to show the robustness of the proposed LA scheme. A small K value indicates higher multi-path interference level, and vice versa. It can be seen from Fig. 4.33a and b that the proposed MCS selection method is robust to different level of multi-path fading interference. When K decreases, i.e., there are more multi-path noises, significant average channel utilities can still be achieved, especially for 2-ray path loss model. In contract, the performance of the statistic-based sampling scheme, i.e., ARF, degrade drastically, since such statistic based LA scheme can be easily interfered by short-term transmission failures. Figure 4.35 shows the proportion of different MCS levels selected by the sampling scheme and the proposed method. It can be observed the MCS selection of our method is more balanced than sampling scheme and similar to the optimal selection. The sampling scheme is sensitive to sudden transmission failure, thus the MCS selections are more distributed on lower levers. The sampling method does not retain the packet in the buffer queue. Instead, it keeps sending packets as soon as there is a transmission opportunity even in severe channel conditions.

4.5.4.5 Resistance to Trajectory Deviation

To verify the feasibility of the proposed MCS selection in practical scenarios, where the maritime vehicle might not travel exactly along the sea lanes, e.g., due to the ocean current and inaccuracy GPS information. The ACU performance with different position deviation is shown in Fig. 4.34a. The deviation is defined as the mean square error of the distance difference between the practical trajectory and the fixed sea lane. It is shown that the proposed method is stable when the position deviation is less than 15 m, and maintain relative ACU around 70% until the deviation is larger than 65 m. Such results show that if the maritime vehicle does not deviate from the scheduled sea lane too much, the proposed method can achieve decent performance.

4.5.4.6 Impact of Trace Set Size X

In practical usage, the historical records trace may not always be available. To find out the number of trace trips X required to train an effective NN, the ACU performance with different sizes of trace sets is shown in Fig. 4.34b. When the sizes X increase from 5 to 11, the ACU performance increases significantly, as the NN can better fit to the SNR variations based on adequate trace data. When the trace size continues to increase, the ACU performance gain is insignificant and cannot increase the accuracy of the NN. From these simulation results, it is demonstrated that, compared with traditional methods, our NARNN prediction based LA scheme can significantly improve the ACU based on a small size NN with limited lag order and training data set size. Besides, the proposed LA scheme also shows robustness to both the channel fast fading from multi-path interference and the position deviation from the sea lanes.

4.5.5 Summary

In this section, we have studied the NARXNN based prediction method for selecting the MCS for maritime vehicles when traveling along fixed sea lanes. The historical SNR records and the transmission distance are used to train and predict the channel conditions for every egress frame from the vehicle to the access station. We have employed different architectures to train and test the performance of the NN, and showed that our proposed scheme can greatly increase the average utility of the wireless channel between the mobile maritime vehicle and the land access station. It is shown that the proposed algorithm is robust given different levels of multi-path channel fading interference as well as a certain level of mobility deviation. Such research results can provide inspirations to design efficient, robust and scalable LA methods for more general mobile Internet of things.

References

1. A. Meola, How the internet of things will transform private and public transportation. http://uk.businessinsider.com/internet-of-things-connected-transportation-2016-10. Accessed 2 Apr 2018
2. J. Ott, D. Kutscher, The Drive-thru architecture: WLAN-based Internet access on the road, in *Proc. IEEE VTC Spring*, vol. 5 (2004), pp. 2615–2622
3. N. Cheng, N. Lu, N. Zhang, X. Shen, J.W. Mark, Opportunistic WiFi offloading in vehicular environment: A queueing analysis, in *IEEE Global Communications Conference (GLOBECOM)* (2014), pp. 211–216
4. H. Wu, W. Xu, J. Chen, L. Wang, X. Shen, Matching-based content caching in heterogeneous vehicular networks, in *Proc. IEEE Global Communications Conference (GLOBECOM)* (IEEE, Piscataway, 2018), pp. 1–6
5. Z. Su, Q. Xu, Y. Hui, M. Wen, S. Guo, A game theoretic approach to parked vehicle assisted content delivery in vehicular ad hoc networks. IEEE Trans. Veh. Technol. **66**(7), 6461–6474 (2017)
6. D. Jiang, V. Taliwal, A. Meier, W. Holfelder, R. Herrtwich, Design of 5.9 GHz DSRC-based vehicular safety communication. IEEE Wirel. Commun. **13**(5), 36–43 (2006)
7. J. Choi, V. Va, N. Gonzalez-Prelcic, R. Daniels, C.R. Bhat, R.W. Heath, Millimeter-wave vehicular communication to support massive automotive sensing. IEEE Commun. Mag. **54**(12), 160–167 (2016)
8. Y. Yao, X. Chen, L. Rao, X. Liu, X. Zhou, LORA: Loss differentiation rate adaptation scheme for vehicle-to-vehicle safety communications. IEEE Trans. Veh. Technol. **66**(3), 2499–2512 (2017)
9. D. Halperin, W. Hu, A. Sheth, D. Wetherall, Predictable 802.11 packet delivery from wireless channel measurements. ACM SIGCOMM Comput. Commun. Rev. **41**(4), 159–170 (2011)
10. X. Cao, R. Ma, L. Liu, H. Shi, Y. Cheng, C. Sun, A machine learning-based algorithm for joint scheduling and power control in wireless networks. IEEE Internet Things J. **5**(6), 4308–4318 (2018)
11. R. GhasemAghaei, M.A. Rahman, W. Gueaieb, A. El Saddik, Ant colony-based reinforcement learning algorithm for routing in wireless sensor networks, in *Proc. IEEE IMTC* (IEEE, Piscataway, 2007), pp. 1–6
12. N. Mastronarde, M. van der Schaar, Fast reinforcement learning for energy-efficient wireless communication. IEEE Trans. Signal Process. **59**(12), 6262–6266 (2011)
13. W. Xu, H.A. Omar, W. Zhuang, X. Shen, Delay analysis of in-vehicle internet access via on-road WiFi access points. IEEE Access **5**, 2736–2746 (2017)
14. W. Xu, H. Zhou, Y. Bi, N. Cheng, X. Shen, L. Thanayankizil, F. Bai, Exploiting hotspot-2.0 for traffic offloading in mobile networks. IEEE Netw. **32**(5), 131–137 (2018)
15. G. Bianchi, Performance analysis of the IEEE 802.11 distributed coordination function. IEEE J. Sel. Areas Commun. **18**(3), 535–547 (2000)
16. R. Combes, J. Ok, A. Proutiere, D. Yun, Y. Yi, Optimal rate sampling in 802.11 systems: Theory, design, and implementation. IEEE Trans. Mobile Comput.. Early access. https://doi.org/10.1109/TMC.2018.2854758
17. P. Group et al., PTV VISSIM. *Retrieved from PTV Group*. http://vision-traffic.ptvgroup.com/en-us/products/ptv-vissim/ (2015)
18. Z. Wang, W. Yan, T. Oates, Time series classification from scratch with deep neural networks: A strong baseline, in *2017 International Joint Conference on Neural Networks (IJCNN)* (IEEE, Piscataway, 2017), pp. 1578–1585
19. L. Bedogni, A. Trotta, M. Di Felice, Y. Gao, X. Zhang, Q. Zhang, F. Malabocchia, L. Bononi, Dynamic adaptive video streaming on heterogeneous TVWS and Wi-Fi networks. IEEE/ACM Trans. Netw. **25**(6), 3253–3266 (2017)

20. A. Gonzalez-Ruiz, A. Ghaffarkhah, Y. Mostofi, A comprehensive overview and characterization of wireless channels for networked robotic and control systems. J. Robot. **2011** 101–119, (2011)
21. X. Cai, G.B. Giannakis, A two-dimensional channel simulation model for shadowing processes. IEEE Trans. Veh. Technol. **52**(6), 1558–1567 (2003)
22. S. Vitturi, L. Seno, F. Tramarin, M. Bertocco, On the rate adaptation techniques of IEEE 802.11 networks for industrial applications. IEEE Trans. Ind. Informat. **9**(1), 198–208 (2012)
23. I. 802.11af 2013, Part 11: Wireless LAN medium access control (MAC) and physical layer (PHY) specifications amendment 5: Television white spaces (TVWS) operation. IEEE Std **802**(11) (2013)
24. M. Lacage, M.H. Manshaei, T. Turletti, IEEE 802.11 rate adaptation: A practical approach, in *Proceedings of the 7th ACM International Symposium on Modeling, Analysis and Simulation of Wireless and Mobile Systems* (ACM, New York, 2004), pp. 126–134
25. K.-L. A. Yau, A.R. Syed, W. Hashim, J. Qadir, C. Wu, N. Hassan, Maritime networking: Bringing internet to the sea. IEEE Access **7**, 48236–48255 (2019)
26. R. Campos, T. Oliveira, N. Cruz, A. Matos, J.M. Almeida, Bluecom+: Cost-effective broadband communications at remote ocean areas, in *OCEANS 2016-Shanghai* (IEEE, Piscataway, 2016), pp. 1–6
27. F.B. Teixeira, T. Oliveira, M. Lopes, J. Ruela, R. Campos, M. Ricardo, Tethered balloons and tv white spaces: A solution for real-time marine data transfer at remote ocean areas, in *2016 IEEE Third Underwater Communications and Networking Conference (UComms)* (IEEE, Piscataway, 2016), pp. 1–5
28. L. Pilosu, A. Autolitano, D. Brevi, R. Scopigno, Exploring tv white spaces for the mitigation of AIS weaknesses, in *2015 IEEE Symposium on Communications and Vehicular Technology in the Benelux (SCVT)* (IEEE, Piscataway, 2015), pp. 1–6
29. S. Akshaya, S.N. Rao, Comparison of long range Wi-Fi and i 802.22 for marine connectivity, in *2017 IEEE International Conference on Computational Intelligence and Computing Research (ICCIC)* (IEEE, Piscataway, 2017), pp. 1–4
30. J. Wang, H. Zhou, Y. Li, Q. Sun, Y. Wu, S. Jin, T.Q. Quek, C. Xu, Wireless channel models for maritime communications. IEEE Access **6**, 68070–68088 (2018)
31. W. Xu, H. Zhou, N. Cheng, F. Lyu, W. Shi, J. Chen, X. Shen, Internet of vehicles in big data era. IEEE/CAA J. Autom. Sinica **5**(1), 19–35 (2018)
32. J. Joo, M.C. Park, D.S. Han, V. Pejovic, Deep learning-based channel prediction in realistic vehicular communications. IEEE Access **7**, 27846–27858 (2019)
33. R.M. Sandoval, A.-J. Garcia-Sanchez, J. Garcia-Haro, Optimizing and updating lora communication parameters: a machine learning approach. IEEE Trans. Netw. Service Manag. **16**, 884 (2019)
34. W. Xu, H. Zhou, H. Wu, F. Lyu, N. Cheng, X. Shen, Intelligent link adaptation in 802.11 vehicular networks: Challenges and solutions. IEEE Commun. Stand. Mag. **3**(1), 12–18 (2019)
35. T. Wei, W. Feng, Y. Chen, C.-X. Wang, N. Ge, J. Lu, Hybrid satellite-terrestrial communication networks for the maritime internet of things: key technologies, opportunities, and challenges. Preprint. arXiv:1903.11814 (2019)
36. Y. Zhang, Z. Chen, F. Dong, B. Chen, Maritime wireless broadband communication system based on TVWS, in *2015 4th International Conference on Mechatronics, Materials, Chemistry and Computer Engineering* (Atlantis Press, 2015)
37. C. Zhang, K. Ota, J. Jia, M. Dong, Breaking the blockage for big data transmission: Gigabit road communication in autonomous vehicles. IEEE Commun. Mag. **56**(6), 152–157 (2018)
38. H. Li, K. Ota, M. Dong, Deep reinforcement scheduling for mobile crowdsensing in fog computing. ACM Trans. Internet Technol. (TOIT) **19**(2), 21 (2019)
39. A.U. Joshi, P. Kulkarni, Vehicular WiFi access and rate adaptation, in *Proc. ACM SigCom* (2010), pp. 423–424
40. D. Hadaller, S. Keshav, T. Brecht, S. Agarwal, Vehicular opportunistic communication under the microscope, in *Proc. ACM Mobile Systems, Applications and Services* (2007), pp. 206–219

41. Z.A. Qazi, S. Nadeem, Z.A. Uzmi, MAC rate adaptation and cross layer behavior for vehicular WiFi access: An experimental study. Preprint. arXiv:1610.03834 (2016)
42. H. Li, K. Ota, M. Dong, Learning IoT in edge: Deep learning for the internet of things with edge computing. IEEE Netw. **32**(1), 96–101 (2018)
43. G. Holland, N. Vaidya, P. Bahl, A rate-adaptive MAC protocol for multi-hop wireless networks, in *Proc. ACM Mobile Computing and Networking* (2001), pp. 236–251
44. G. Judd, X. Wang, P. Steenkiste, Efficient channel-aware rate adaptation in dynamic environments, in *Proc. ACM Mobile Systems, Applications, and Services* (2008), pp. 118–131
45. M.O. Khan, L. Qiu, Accurate WiFi packet delivery rate estimation and applications, in *Proc. IEEE INFOCOM* (2016), pp. 1–9
46. Z. Zhou, X. Chen, E. Li, L. Zeng, K. Luo, J. Zhang, Edge intelligence: Paving the last mile of artificial intelligence with edge computing. Preprint. arXiv:1905.10083 (2019)
47. X. Chen, Q. Shi, L. Yang, J. Xu, Thriftyedge: Resource-efficient edge computing for intelligent IoT applications. IEEE Netw. **32**(1), 61–65 (2018)
48. R.T. Fleifel, S.S. Soliman, W. Hamouda, A. Badawi, LTE primary user modeling using a hybrid ARIMA/NARX neural network model in CR, in *2017 IEEE Wireless Communications and Networking Conference (WCNC)* (IEEE, Piscataway, 2017), pp. 1–6
49. V.E. Narawade, U.D. Kolekar, NNRA-CAC: NARX neural network-based rate adjustment for congestion avoidance and control in wireless sensor networks. New Rev. Inf. Netw. **22**(2), 85–110 (2017)
50. Z. Boussaada, O. Curea, A. Remaci, H. Camblong, N. Mrabet Bellaaj, A nonlinear autoregressive exogenous (NARX) neural network model for the prediction of the daily direct solar radiation. Energies **11**(3), 620 (2018)
51. Y.S. Meng, Y.H. Lee, Measurements and characterizations of air-to-ground channel over sea surface at c-band with low airborne altitudes. IEEE Trans. Veh. Technol. **60**(4), 1943–1948 (2011)
52. J.-H. Lee, J. Choi, W.-H. Lee, J.-W. Choi, S.-C. Kim, Measurement and analysis on land-to-ship offshore wireless channel in 2.4 GHz. IEEE Wireless Commun. Lett. **6**(2), 222–225 (2017)
53. C.R. Stevenson, G. Chouinard, Z. Lei, W. Hu, S.J. Shellhammer, W. Caldwell, IEEE 802.22: The first cognitive radio wireless regional area network standard. IEEE Commun. Mag. **47**(1), 130–138 (2009)
54. A. Goldsmith, *Wireless Communications* (Cambridge University Press, Cambridge, 2005)
55. A. Graves, A.-r. Mohamed, G. Hinton, Speech recognition with deep recurrent neural networks, in *2013 IEEE International Conference on Acoustics, Speech and Signal Processing* (IEEE, Piscataway, 2013), pp. 6645–6649
56. I. Sutskever, O. Vinyals, Q.V. Le, Sequence to sequence learning with neural networks, in *Advances in Neural Information Processing Systems* (2014), pp. 3104–3112
57. Y. Li, R. Yu, C. Shahabi, Y. Liu, Diffusion convolutional recurrent neural network: Data-driven traffic forecasting. Preprint. arXiv:1707.01926 (2017)
58. J.J. Moré, The Levenberg-Marquardt algorithm: implementation and theory, in *Numerical Analysis* (Springer, Berlin, 1978), pp. 105–116
59. shipxy, “shipxy services”. http://a.shipxy.com/. Accessed 25 Oct 2019

Chapter 5
Intelligent Networking enabled Vehicular Distributed Learning

Mobile computing has emerged as an important paradigm to envision the 'last mile' of computing services to mobile users. There are many novel distributed computing methods which can be applied for vehicle users to let them cooperatively train ML models for future AI applications, However, traditional centralized training methods are not suitable for vehicle users since they are not connected by a reliable and bandwidth-rich Internet access, which is highly dynamic and often suffers a lot from interruption, interference, etc. In this chapter, we exploit two learning paradigms and analyze their performance based on the IoV connectivity and exploit the vehicles' mobility. Due to the specific mobility pattern and communication characteristics in IoV, excessive training latency can be caused by the communication bandwidth constraints in vehicular environments, non-negligible volumes of iteration parameters and heterogeneity in computing capacities of distributed workers, etc. We propose novel computing methods to seek the possibility to provision artificial intelligence (AI) to the mobility world with the help of IoV, which has great potential to bring the power of AI to all road users, to support a variety of intelligent applications, e.g., autonomous driving, road safety, ITS, etc.

5.1 Background and Motivation

Deploying ML models on road can help to provision ubiquitous AI services to mobile users, where a roadside aggregation server and workers within its coverage area can cooperatively train models to leverage the distributed big data and computation resources [1, 2]. As vehicle users are often stay at the edge of network, edge computing are considered to offload the computing tasks from cloud center to edge stations while reduce the communication overhead to upload all raw data [3]. Federated learning (FL) is considered as the most promising technology for edge users, which keeps the raw data locally and exchange the gradients parameters only

W. Xu et al., *Internet Access in Vehicular Networks*,
https://doi.org/10.1007/978-3-030-88991-3_5

between workers and the edge server, and thus can boost the training process and protect the user privacy. Such benefits are essential for future intelligent services, e.g., location-aware applications, autonomous driving, etc. [4–6] On the other hand, machine learning models can also help to improve the IoV connectivity, e.g., improve the intelligent access and new communication paradigms [7, 8].

In order to efficiently train neural models for vehicle users, many literature have been proposed to adapt the aggregation iteration procedure for heterogeneous computing conditions or reduce the communication bandwidth consumption. In [9, 10], the author proposed asynchronous decentralized algorithm to train ML models, and show that it can achieve comparable convergence rate as synchronous centralized counterparts. Asynchronous mechanism can well adapt to the highly heterogeneous configurations edge users, which avoid the waiting time for parameters synchronization, e.g., due to stragglers. There are many extensions based on the asynchronous settings in edge environment. In [11], the authors propose to do trade-off between local training and global aggregation. Zhang et al. consider the staleness issue in the asynchronous settings and adjust the learning rate to count the staleness from the heterogeneous platforms [12]. From the communication optimization perspective, Reisizadeh et al. propose to apply periodic averaging and do quantization to do the trade-off between communication and computation [13]. Amiri et al. utilize the additive nature of analog channel to do the parameters aggregation over-the-air, whereby the bandwidth consumption for the upload gradients can be significantly reduced [14]. However, such method requires communication synchronization for distributed workers when uploading the gradients, and is not compatible with current vehicular communication system.

In real practice, vehicles big raw data including OBU sensing status, camera streaming, etc., are of extremely large volume and various types. To utilize such big data, it is unpractical to upload them to a global server to conduct centralized ML training. In this chapter, we consider to keep the raw data within the vehicle and conduct distributed learning among multiple vehicle users, who will cooperative training ML models with an RSU. We propose two method to optimize the learning process in mobile conditions via the interplay of IoV and asynchronous parallelization scheme. We first combine the FL with IoV with carefully designed communication protocol, then we consider the vehicle as a carrier of the learning model. By interaction between vehicles and the roadside server, and employ asynchronous parallelization scheme to boost the learning process. It is shown that the proposed methods can achieve significant learning performance with much reduce communication cost for vehicle users.

5.2 Rateless Coding Enabled Broadcasting for Vehicular Federated Learning

The motivation of this section is to boost the distributed training process for vehicle users by taking the advantage of asynchronous parallelization and the property of the edge-to-vehicles broadcasting channel, i.e., linear growth with the number of users. Compare with full asynchronous parallelization where both the parameters upload to edge server and the updated model delivery to workers fully follows asynchronous manner, we employ the asynchronous parameters uploading while in the downlink broadcast the updated model via the wireless channel in synchronous way. The advantages are two-fold. First, asynchronous parameters model aggregation can overcome the straggler problem caused by the device heterogeneity and highly dynamic communication channel in edge environment. Secondly, model broadcasting via wireless channel can significantly reduce the time for delivering the updated model to users, since all users would simultaneously receive a copy from the edge station given perfect channel status.

The way of model broadcasting using current IoV architecture is to partitions the model file into a set of packets, e.g., UDP fragmentation, and then send to air over IoV interface, often with the most conservative modulation and coding scheme (MCS) to reduce the packet error probability. Normally there is no ACK frame for broadcast packets, and if there is any packet drop due to the channel loss, the entire broadcast process needs to be repeated since it is difficult for the edge station to identify the dropped packets for multiple users. It is common in edge environment, especially for mobile users, that the communication to the edge station would be interrupted due to channel fading, mobility, etc., and thus the time for delivering the updated model in each round would consume significant time. Per our observation, the communication time for model broadcasting is apparently non-negligible comparing with the local training duration. In order to optimize the broadcast time overhead, we employ the rateless code to generate a set of encoded packets from the model file in each round, which can be decoded for any user who has received enough number encoded packets from the set. Increasing the broadcasting time for the encoded packets may help to cover more vehicle users, however may increase the time consumption in each iteration round. Our mechanism is to tune the broadcasting time for these encoded packets, and thus the proportion of the users that received the updated model can be controlled. Specifically, we apply a multi-armed bandit (MAB) algorithm to intelligent adapt the broadcasting time to optimize the training procedure.

In this section, we propose to boost the distributed learning via a novel broadcasting enabled asynchronous parallelization (BAP) scheme in mobile edge environment, where the parameters are aggregated asynchronously, and the updated model are dispatched to distributed workers via the one-hop broadcasting channel in each round. Specifically, the edge server aggregates the received gradients with a first-in-fist-out (FIFO) manner to avoid the waiting time for stragglers, and utilize the rateless coding based model broadcasting to accelerate the downlink parameters delivery to mobile workers, which is conducted by intelligently tuning the broadcasting time in each round. We provide analytical proof to show that the convergence rate of BAP is comparable to traditional synchronous algorithms, and demonstrate that BAP can improve the convergence speed by 30% over the statue-of-the-art asynchronous methods from extensive empirical experiments.

5.2.1 System Model

We consider a road area where there are several blocks and intersection roads, as shown in Fig. 5.1. A roadside AP is located at the central position of the map, who is served as an aggregation server (AS). Vehicles will conduct local training for K round and then upload the gradients to the AS, who will conduct the aggregation and then broadcast the updated model to the air. Before the real transmission, the model file will be encoded using the raptor code, which will output a set of encoded data packets. These files will be transmitted to the broadcast channel to cover a set of vehicles. The longer the broadcast time, the higher the number of vehicles that can receive the updated model parameters in this round.

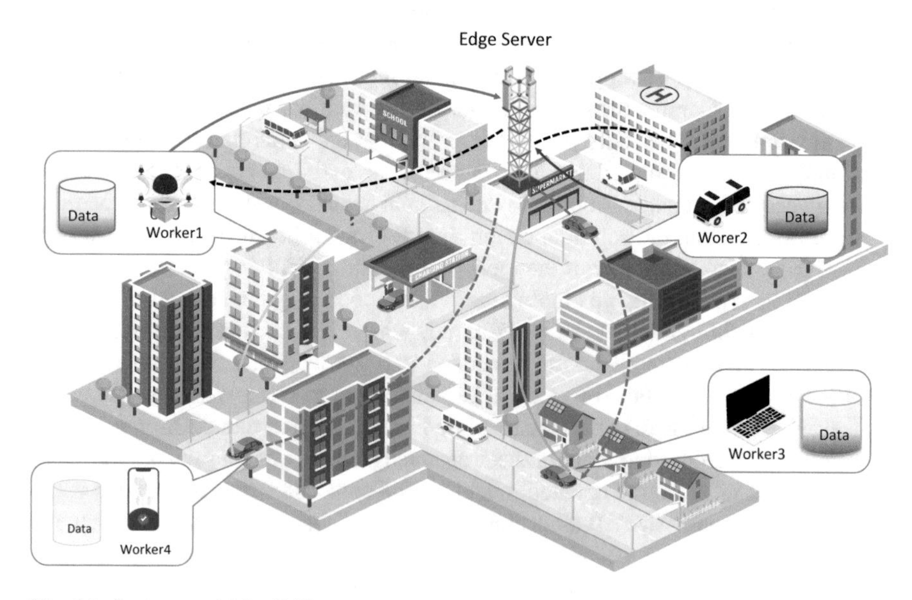

Fig. 5.1 System model for BAP

5.2.2 BAP for Vehicular Cooperative Learning

We consider an IoV scenario where n vehicles are cooperatively training a shared model with a roadside server. For each vehicle $i \in [n]$, there exists a local database D_i. The overall goal is to train the global model $w \in \mathbf{R}^d$ over the distributed datasets from all vehicles, which are located in the AS. The wireless connection between each user and the AS suffers from various channel fading, e.g., multi-path, shadowing, user mobility, etc. Denote the loss function on data sample z by $f(w; z)$, then we define the overall cost function as

$$F(w) = \frac{1}{n} \sum_{i=1}^{n} \mathbf{E}_{z^i \sim D^i} f(w; z^i) \tag{5.1}$$

And thus the optimization problem can be formulated as

$$\min_{w \in \mathbf{R}^d} F(w), \tag{5.2}$$

Consider a single execution in each round with P global epochs. In the t-th epoch, the AS receives an accumulated gradient G_t^i from an arbitrary vehicle $i \in [n]$, and updates the global model as following:

$$w_{t+1} = w_t + G_t^i \tag{5.3}$$

Once the global model is updated, the AS immediately broadcast new model to some n vehicles depends on the broadcast time. On arbitrary device i, vehicles uploads accumulate gradient G_t^i for every K iterations, which contributes to global model w_{t+1} in (5.3). G_t^i is calculated by

$$G_t^i = \sum_{k=0}^{K-1} \nabla f(w_{t,k-1}^i, z_{t,k}^i) \tag{5.4}$$

where $z_{t,k}^i \sim D^i$ denotes the sample randomly selected in local dataset D^i. In iteration $k \in [K]$, local model $w_{t,k}^i$ is defined as

$$w_{t,k}^i = \begin{cases} w_\tau & \text{receive } w_\tau \text{ from server} \\ w_{t,k-1}^i - \gamma \nabla f(w_{t,k-1}^i, z_{t,k}^i) & \text{otherwise} \end{cases} \tag{5.5}$$

This means the local model $w_{t,k}^i$ would be replaced by w_τ if vehicle receives broadcast of new global model in that iteration. Otherwise, $w_{t,k}^i$ would be derived from $w_{t,k-1}^i$. The algorithm workflow is summarized in Algorithm 3 and demonstrated in Fig. 5.2.

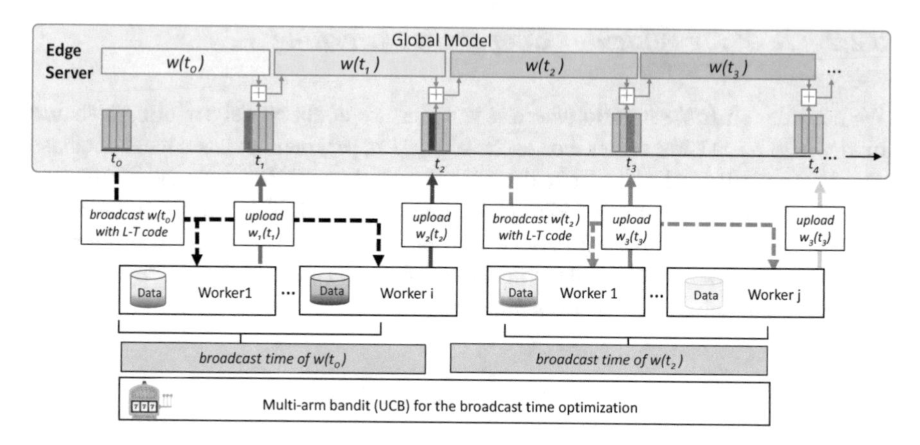

Fig. 5.2 Demonstration of the BAP flow

We employ the raptor code [15, 16] to encode the global model if there is an update from any vehicle. The encoder will continuously output coded packets with a size of s bytes from a model file of size m. As stated in Algorithm 3, the server will control the broadcast time, i.e., the coded packets, which will lead to different number of vehicles that can successfully recover the model file by doing a tradeoff between the communication time overhead and the training efficiency from more workers. We apply the conventional Upper Confidence Bound (UCB1) algorithm (a category of the Multi-Armed Bandit solutions) to deduce the best broadcast time in each FL round, to strike a balance between the exploration vs exploitation dilemma [17]. Short broadcast time would reduce the ratio of the workers that can download the updated global model, while in contract long broadcast time can improve the number of workers that keep synchronized with the AS. The broadcast time is discretized into a set of actions, i.e., the action set. Each round one action is selected from the action set, and the actual reward is used to evaluate the selected action iteratively.

5.2.3 *Convergence Analysis*

In this section, we simplify $f(w, z)$ as $f(w)$ for loss function with sampling. To analyze the convergence rate of BAP, we adopt a different scheme to describe the convergence step of the algorithm. Consider a training process with P iterations, we maintain N sequences $\{w_r^i\}_{r=0}^P$ for each worker $i \in [N]$ and evaluate the convergence rate, which is defined as

$$w_r = \frac{1}{N}\sum_{i=1}^{N} \nabla f(w_r^i) \tag{5.6}$$

Algorithm 3 BAP

Input: Number of vehicles n, learning rate γ
Initialize: $\boldsymbol{\omega}_1$, $isPull = False$;
For each vehicle $i = 1, \ldots, n$:
Simultaneously run these two threads:
Receiving Thread:
  repeat:
    Receive encoded packets from server, if can decode, then recover the global model ω_t from server;
    Set the pulled flag $isPull = True$;
  until loss reaches threshold
Local Training Thread:
  repeat:
    Initialize $G^i_{t,0} = 0$;
    if $isPull$ is $True$ then
      Replacing the vehicle model by the received one $\boldsymbol{\omega}^i_{t,0} = \boldsymbol{\omega}_\tau$;
      $isPull = False$;
    end if
    for $k = 0$ to $K - 1$:
      if $isPull$ is $True$ then
        Replacing the vehicle model by the received one $\boldsymbol{\omega}^i_{t,k} = \boldsymbol{\omega}_\tau$;
        $isPull = False$;
      end if
      Randomly sample a mini-batch $z^i_{t,k}$;
      Compute the stochastic gradient $\nabla f_i(\boldsymbol{\omega}_{t,k}, z^i_{t,k})$;
      Update parameters $\boldsymbol{\omega}^i_{t,k+1} = \boldsymbol{\omega}^i_{t,k} - \gamma \nabla f_i(\boldsymbol{\omega}_{t,k}, z^i_{t,k})$;
      Accumulate $G^i_{t,k+1} = G^i_{t,k} + \nabla f_i(\boldsymbol{\omega}_{t,k}, z^i_{t,k})$;
    end for
    Send $G^i_{t,K}$ to server;
  until Convergence
On server:
repeat:
  Receive G^i_t from any worker i;
  Update global model $\omega_{t+1} = \omega_t - \gamma G^i_t$;
  Encode the updated model ω_{t+1} into a set of packets, repeatedly broadcast all of these packets to all workers for t time;
until Convergence

We also denote $r^i_0 \leq r$ for the latest iteration that worker i receives the broadcast. The update process of the local model can thus be described in the following way:

$$w^i_r = \begin{cases} w^i_{r^i_0} & r^i_0 = r \\ w^i_{r-1} - \gamma \nabla f(w^i_{r-1}) & r^i_0 < r \end{cases} \tag{5.7}$$

where

$$w_r^i = w_0 - \frac{\gamma}{N} \sum_{h=1}^{N} \sum_{j=0}^{S_r^{i,h}} \nabla f(w_j^h) \tag{5.8}$$

We denote w_0 as the initial parameter for all workers, and $S_r^{i,h}$ as the latest iteration of worker h that has written the gradient to global model when worker i receives broadcast at local iteration r. We define delay T and broadcast interval e so that $r - r_0^i \leq e$ and $|r - S_r^{i,h}| \leq T$ for all $r \in [P]$ and $i, h \in [N]$. Note that $e \sim O(\frac{K}{N})$ since broadcast becomes more frequently as worker size increases.

We also take additional assumptions as following:

1. **Lipschitzian Gradient:** The gradient function $\nabla F(w)$ of the objective function f is Lipschitz continuous with constant $L > 0$, where

$$||\nabla F(w) - \nabla F(w')||^2 \leq L^2 ||w - w'||^2 \tag{5.9}$$

2. **Bounded Variance:** there exists scalar $\sigma \geq 0$ such that, for all $w \in \mathbb{R}^d$,

$$||\nabla f(w) - \nabla F(w)||^2 \leq \sigma^2 \tag{5.10}$$

3. **Second Moment:** there exists scalar $G \geq 0$ such that, for all $w \in \mathbb{R}^d$,

$$||\nabla f(w)||^2 \leq G^2 \tag{5.11}$$

By denoting the minimum value of the loss function as $F^\star$, we establish the following theorem for the convergence of APSB.

Theorem 5.1 *Consider problem (5.2) under the above assumption. When the learning rate satisfies*

$$\gamma L < \frac{1}{4} \tag{5.12}$$

we have

$$\frac{1}{2S_P} \sum_{r=1}^{P} a_r (F(w_r) - F^*)$$

$$\leq \frac{\mu b^3}{4S_P} w_0 + \frac{2P(P+2b)\sigma^2}{\mu N S_P} + \frac{48(2e^2 + T^2)PG^2}{\mu^2 S_P} \tag{5.13}$$

where sequence $a_r = (b+r)^2$, $S_P = \frac{P}{6}(2P^2 + 6bP - 3P + 6b^2 - 6b + 1) \geq \frac{1}{3}P^3$ and constants $b \geq 0$

We optimize the convergence factor of $\frac{96(2e^2+T^2)PG^2L}{\mu^2 S_P}$ from $\frac{256P(K+T)^2G^2}{\mu^2 S_P}$ in [18], where $e \sim O(K/N)$

5.2.4 *Performance Evaluation*

In this section, we combine the vehicle mobility with the asynchronous aggregation process to evaluate our BAP scheme. We consider a map which include 4×4 blocks as shown in Fig. 5.3, where n vehicles are driving within the area with the velocity range given in Table 5.1, who collect the raw data and conduct local training and interact with the server. The experiment parameters including the communication channel fading and road configuration are listed in Table 5.1.

We have utilized multiple vehicles to train a neural network, i.e., ResNet [19]. Figure 5.4 shows the training procedure, i.e., the loss value vs. the time, which include both the computing and communication duration. The baseline include the following cases, all of them are employing the peer-to-peer communication, i.e., the server will start an independent thread to transmit the whole model to all vehicles, who would then perform the local training and upload the local model to the AS.

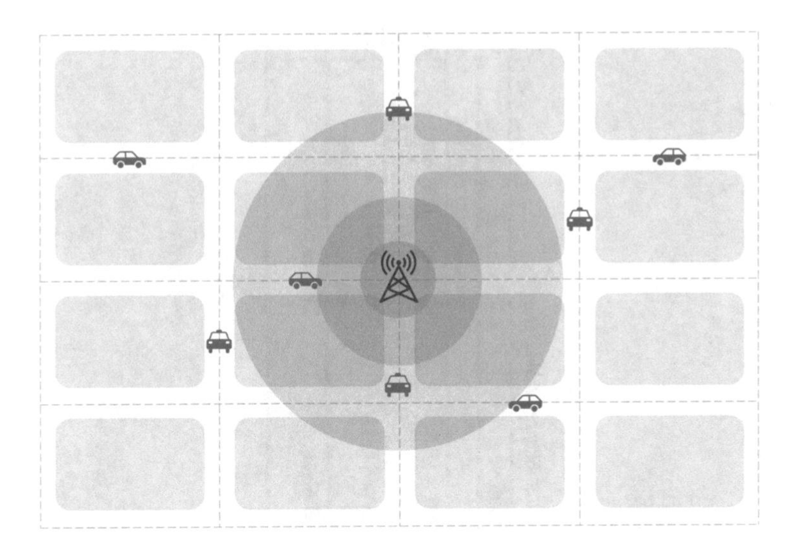

Fig. 5.3 Map of the IoV scenario for BAP

Table 5.1 Experiment parameter for BAP

Parameter	Value	Parameter	Value	Parameter	Value
Number of vehicles	32	Power of shadowing	12 dB	Broadcast time selection	UCB1 [17]
Block size	$80 \times 80\,\mathrm{m}^2$	Transmit frequency	2.4 GHz	Protocol	802.11n, 20 MHz
Street width	24 m	Fast fading noise (dB)	15	WiFi power	25 dbm
Vehicle velocity	30–60 km/h	Path loss constant/exponent	−12.89 dB/3	Codec	Raptor code
ML model	ResNet	Noise power	−85 dbm	Encoded packet size	2300 Byte

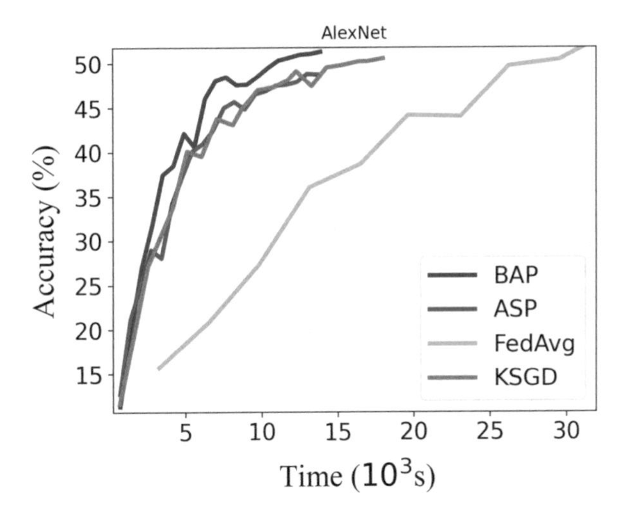

Fig. 5.4 Training Loss of BAP and baseline methods over Alex model

1. KSGD: Kalman-based stochastic gradient descent using normal broadcast [20], i.e., when the server updates the model, it will broadcast the model file using UDP protocol until all workers received it by checking an ACK from each vehicle.
2. FedAvg: the traditional federated learning method that employ synchronous aggregation from all vehicles. The updated global model will be broadcast to all users until received [21].
3. ASP: asynchronous parallel scheme that both the vehicle uplink push and global model downlink is fully asynchronous [9]. Vehicle and the server has independent communication link to transfer both the parameters and the updated model file.

As shown in Fig. 5.4, our proposed BAP is the most efficient in the distributed training for the AlexNet model, which is of small size and normally are easy to train. It is shown that BAP can achieve 50% faster in the training speed than the FedAvg. And in terms of the comparison with ASP and KSGD, the BAP scheme also outperform them a lot.

Figure 5.5 demonstrates that for complex model like MNist, the BAP method can outperform the baseline, especially in the beginning stage of the model training. It is shown that to achieve a model accuracy around 90%, the BAP scheme consumes at least 51% less time than the other baselines.

From the results, we can see that our proposed method can efficiently conduct the asynchronous training among mobile vehicle users. It is also shown that proper broadcasting time can be chosen that minimized the overall time used for the model aggregation and achieve the fastest convergence than the baselines.

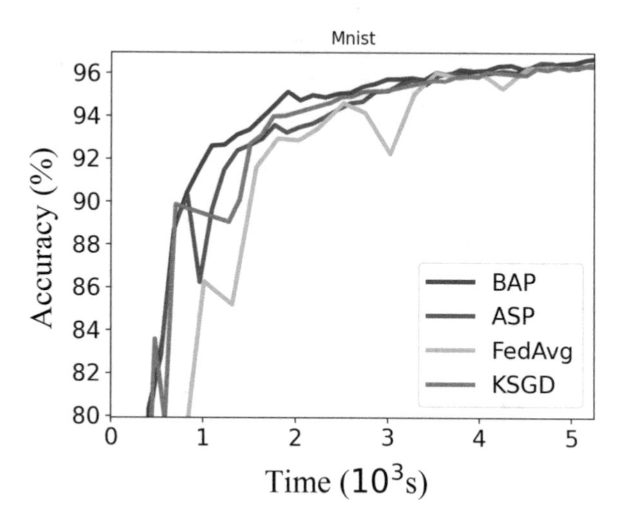

Fig. 5.5 Training Loss of BAP and baseline methods over Mnist Model

5.2.5 Summary

In this section, we have proposed a novel asynchronous parallelization algorithm among mobile vehicles to boost the distributed training in automotive scenario. The rateless code based broadcasting is applied to intelligently tune the mobile workers that can receive the updated model in each round. We have shown that the proposed BAP method can significantly boost the training process, which is crucial to the application of machine learning models in vehicular scenarios.

5.3 Opportunistic Collaborated Learning Over Intelligent Internet of Vehicles

In this section, we consider the scenario where vehicles can interact with two RSUs at each side of a road section. For on-road applications, most of the user data are generated in the driving duration, e.g., the autonomous driving data, road situation detection, etc. Such data are highly determined by the environment during a certain road, which is referred to as the road-related data. The road-related data often contain sensitive user information, e.g., driving behavior, camera data, etc. Besides, the data are often of huge volume. So it is not practical to let vehicle users to upload the data to a could server, which would cause potential privacy leakage and huge communication cost. In addition, keeping such big data can also exhaust the storage space of a vehicle. A feasible way is to conduct in-situ learning while the vehicle is driving through an area, and transmit the learning outcome, e.g., the gradient information to an RSU. Such that there is no need to keep the raw data, while accumulate the experience from multiple vehicles' local learning for training effective ML models. We propose a novel aggregation method that without the backhaul connection, two RSUs can collaboratively train ML models with the vehicle driving through the area via the opportunistic V2R communication (VeOCL). We provide convergence analysis and conduct experiment to show the proposed method's performance.

5.3.1 System Model

As shown in Fig. 5.6 we consider a basic case that include two RSUs at the two side of a road section. We assume that the two RSU are not backhaul connected, so they cannot communicate with each other directly. Such assumption is based on the reality that deploying backhaul connection may cause prohibitive cost, while put an AP at a certain position of road does not need network infrastructure investment. The purpose is to utilize the mobility of vehicles traveling through the road section, i.e., treat the vehicle as a carrier of a model that can be updated via both the local training of the vehicle and the accumulative aggregation from the RSU.

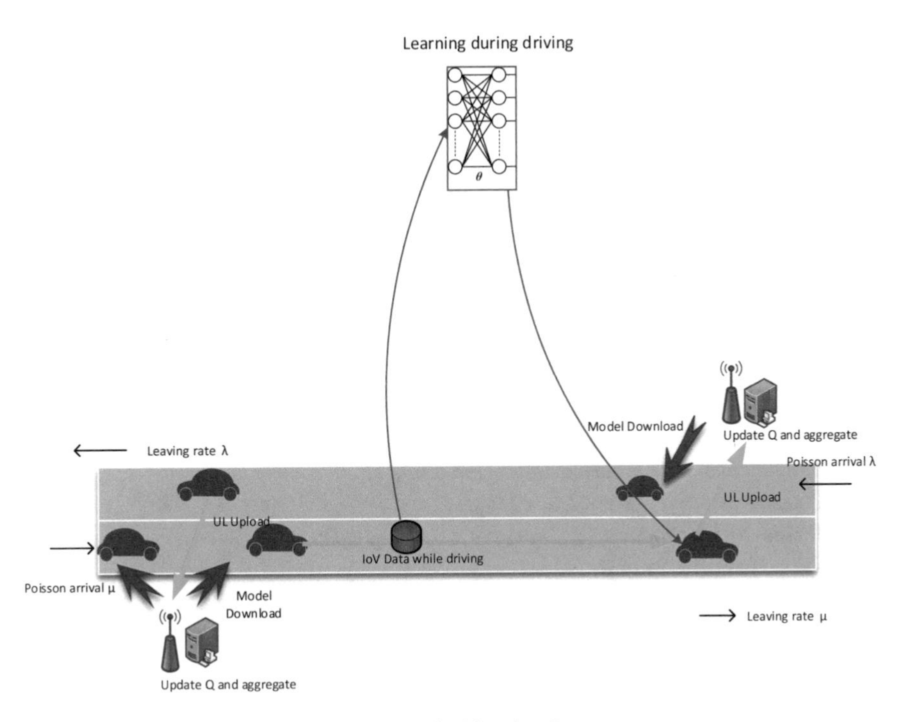

Fig. 5.6 Throughput of different RA schemes in drive-thru Internet

5.3.2 Opportunistic Collaborated Learning via V2R Interaction

We focus on the sum optimization objective for ML model training. Denote the loss value by $f_i(\omega)$ given the model parameters ω for the ith data sample. The objective function is to minimize the loss function by training the model:

$$\min\ F(\omega) = \frac{1}{n}\sum_{i=1}^{n} f_i(\omega), n \text{ is the number of data samples} \tag{5.14}$$

Normally the gradient descent method is applied to optimize the above objective function. Specifically, we employ the stochastic gradient descent (SGD) to iteratively update the model parameters:

$$\omega_{t+1} = \omega_t - \eta_t \mathbf{g}(\omega_t; \xi_t), \tag{5.15}$$

The steps to collaboratively update the ML model is shown in Algorithm 4.

Algorithm 4 Learning among vehicles

Input: Learning rate η
Initialize: $\mathbf{x}_1^1 = \mathbf{x}_1^2 \leftarrow \mathbf{x}$;
On RSU $i = 1, 2$:
Parallel the following three threads:
Sender:
Send the latest local model $\mathbf{x}_t^i$ to vehicle;
Receiver:
Receive model $\mathbf{x}_{t,K_t^j}^j$, $j \neq i$ from vehicle;
Put the received model $\mathbf{x}_{t,K_t^j}^j$ into the queue Q;
Updater:
for $t = 1$ **to** T **do**
  Take the model $\mathbf{x}_{t,K_t^j}^j$, $j \neq i$ out from queue Q;
  Update local model $\mathbf{x}_{t+1}^i \leftarrow \frac{1}{2}(\mathbf{x}_t^i + \mathbf{x}_{t,K_t^j}^j)$;
On vehicle $t = 1, \ldots, N$:
Pull $\mathbf{x}_t^i$ from the nearest station i;
for $k = 0$ **to** $K_t^i - 1$ **do**
  Sample a mini-batch from the road-related data $\xi_{t,k}^i$;
  Compute the stochastic gradient $\nabla F(\mathbf{x}_{t,k}^i, \xi_k^i)$;
  Update local parameter $\mathbf{x}_{t,k+1}^i \leftarrow \mathbf{x}_{t,k}^i - \eta_t \nabla F(\mathbf{x}_{t,k}^i, \xi_{i,k})$;
Push $\mathbf{x}_{t,K_t^i}^i$ to another RSU j;

5.3.3 Convergence Analysis

Similar to previous studies [22], we make the following assumptions, which are widely adopted for analyzing the convergence for distributed learning.

1. **Lipschitz smooth**. The objective function $F : \mathbb{R}^d \rightarrow \mathbb{R}$ is continuously differentiable and the gradient function of F is Lipschitz continuous with Lipschitz constant $L > 0$, i.e.,

$$\|\nabla F(\mathbf{x}) - \nabla F(\tilde{\mathbf{x}})\|_2 \leq L\|\mathbf{x} - \tilde{\mathbf{x}}\|_2$$

   for all $\mathbf{x}, \tilde{\mathbf{x}} \in \mathbb{R}$;
2. **Bounded loss**. The sequence of iterations $\mathbf{x}_t$ is contained in an open set over which F is bounded below by a scalar F_{inf};
3. **Unbiased gradient with bounded variance**. The stochastic gradient $\nabla F(\mathbf{x}; \xi)$ computed from random samples ξ is unbiased for every parameter ω, i.e.,

$$\mathbb{E}_{\xi}[\nabla F(\mathbf{x}; \xi)] = \nabla F(\omega).$$

   The variance of stochastic gradient is bounded

$$\mathbb{E}_{\xi}(\|\nabla F(\mathbf{x}; \xi) - \nabla F(\omega)\|^2) \leq \sigma^2$$

   where σ^2 is a constant.
4. **Initial model**. For simplicity of proof, the value of the model parameter $\mathbf{x}_0$ is initialized to be $\mathbf{0}$;
5. **Vehicle Computing Speed**. We assume the number of local iterations K_t^i in vehicle i follows the exponential distribution with the rate parameter being K for both two stations, i.e., $\mathbb{E}K_t^i = K$ and $Var(K_t^i) = K^2 - \frac{1}{K}$.

The convergence property of VeOCL is give as follows.

Theorem 5.2 (VehicleLearning, Nonconvex Objective, Fixed Stepsize) *Suppose algorithm runs with fixed learning rate $\eta_t = \bar{\eta}$ satisfying*

$$\bar{\eta} \leq \frac{-LK_t^i(1-\rho)^2 + (1-\rho)LA}{4L^2K_t^i(K_t^1 + K_t^2) + 8L^2(1-\rho)^2}, \tag{5.16}$$

where $A = \sqrt{(K_t^i)^2(1-\rho)^2 + 16K_t^i(K_t^1 + K_t^2) + 4(1-\rho)^2}$, for any $i = 1, 2$.

The expected average squared gradient norms of F satisfy the following bound for all $T \in \mathbb{N}$:

$$\frac{1}{T}\sum_{t=1}^{T}\mathbb{E}\|\nabla F(\overline{\mathbf{x}}_t)\|_2^2$$

$$\leq \frac{4(F(\overline{\mathbf{x}}_1) - F_*)}{\bar{\eta}KT} + \frac{L\sigma^2\bar{\eta}}{2} + \frac{L^2\sigma^2\bar{\eta}^2(2K^3 - 1)}{K^2(1-\rho^2)}. \tag{5.17}$$

From Eq. (5.17), we know that VehicleLearning converges to a non-zero constant under a fixed learning rate as $N \to \infty$. The final non-zero constant mainly comes from the second term because the first term being the initial distance approaches to zero gradually. As we can see, the second term is unrelated to the number of iterations N and it diminishes with the learning rate η. To show the relationship between the convergence result and the number of iterations, we have the following corollary.

Corollary 5.1 *Under the condition of Theorem 5.2, if we set*

$$\bar{\eta} = \sqrt{\frac{(F(\omega_1) - F(\omega_\star))P}{\bar{K}L\sigma^2 N}}, \tag{5.18}$$

then for any iteration number

$$N \geq Max(A_2, A_4), \tag{5.19}$$

the output of Algorithm 4 satisfies the following ergodic convergence rate

$$\frac{1}{N}\sum_{t=1}^{N} \mathbb{E}\|\nabla F(\omega_t)\|_2^2 \leq 4\sqrt{\frac{(F(\omega_1) - F(\omega_\star))L\sigma^2}{\bar{K}P}} * \frac{1}{\sqrt{N}}, \tag{5.20}$$

where

$$\begin{aligned}
A_1 &= \frac{3LM}{2} + \frac{3L\bar{K}^2}{2} + 3L\gamma^2(\bar{K}^2 + M) - \frac{3L\bar{K}}{2} + \frac{3L\bar{K}^2}{P}, \\
A_2 &= \frac{2(F(\omega_1) - F(\omega_\star))A_1^2 P^3}{\bar{K}^3 L\sigma^2}, \\
A_3 &= 3K_{max}/2 + 3\bar{K} + 3\gamma^2 K_{max} - 3/2, \\
A_4 &= \frac{K_{max}P(F(\omega_1) - F(\omega_\star))}{2\bar{K}\sigma^2} \\
&\quad \cdot \left(\sqrt{LK_{max}} + \sqrt{LK_{max} + 4A_3}\right)^2.
\end{aligned}$$

Corollary 5.1 shows that the average squared gradient converges to zero with a speed $O(\frac{1}{\sqrt{NP}})$ as the learning rate diminishes. The convergence rate of *LOSP* is in the same order as the sequential SGD when the iteration number N is sufficiently large. Besides, such convergence speed also indicates a linear speedup in terms of the number of workers. Though the convergence rate of *LOSP* is similar to that of *OSP*, its performance has a significant improvement empirically.

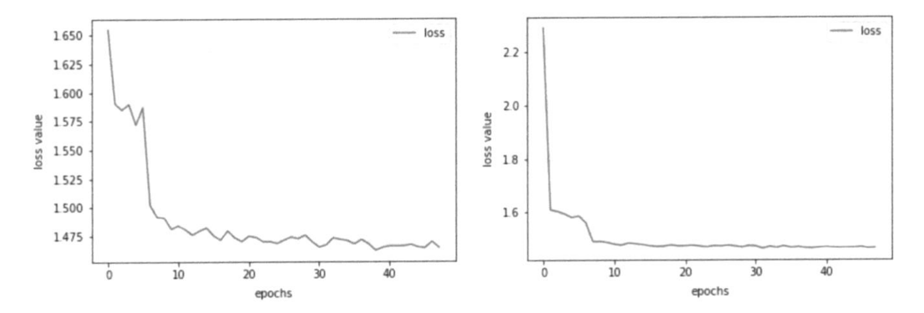

Fig. 5.7 Loss at the two RSUs

We have conducted the experimental evaluation of the proposed model update method, and the results in Fig. 5.7 shows that for such basic case, the model loss value can be decreased significantly in several epochs for both of the two RSUs.

5.3.4 *Summary*

We have conducted an experimental research to show that the loss function at the two RSU side in Fig. 5.7. It is shown that both the two RSU's loss can be optimized quickly, which show that the VeOCL can let different aggregation servers, i.e., the RSU, to efficiently train machine learning models with the vehicles via the opportunistic V2R connection, that does not require to upload the raw data to edge or cloud server. And there is no need for the two RSU's to have the backhaul connection to Internet or each other.

References

1. S. Deng, H. Zhao, W. Fang, J. Yin, S. Dustdar, A.Y. Zomaya, Edge intelligence: the confluence of edge computing and artificial intelligence. IEEE Internet Things J. **7**(8), 7457–7469 (2020)
2. Z. Zhou, X. Chen, E. Li, L. Zeng, K. Luo, J. Zhang, Edge intelligence: paving the last mile of artificial intelligence with edge computing (2019). arXiv preprint arXiv:1905.10083
3. W. Shi, J. Cao, Q. Zhang, Y. Li, L. Xu, Edge computing: vision and challenges. IEEE Internet Things J. **3**(5), 637–646 (2016)
4. W.Y.B. Lim, N.C. Luong, D.T. Hoang, Y. Jiao, Y.-C. Liang, Q. Yang, D. Niyato, C. Miao, Federated learning in mobile edge networks: a comprehensive survey. IEEE Commun. Surv. Tutorials **22**(3), 2031–2063 (2020)
5. H. Zhou, W. Xu, J. Chen, W. Wang, Evolutionary v2x technologies toward the internet of vehicles: challenges and opportunities. Proc. IEEE **108**(2), 308–323 (2020)
6. Y. Wu, L.P. Qian, H. Mao, X. Yang, H. Zhou, X. Tan, D.H. Tsang, Secrecy-driven resource management for vehicular computation offloading networks. IEEE Netw. **32**(3), 84–91 (2018)

7. L. Qian, Y. Wu, N. Yu, F. Jiang, H. Zhou, T.Q. Quek, Learning driven NOMA assisted vehicular edge computing via underlay spectrum sharing. IEEE Trans. Veh. Technol., **70**(1), 977–992 (2021)
8. L.P. Qian, Y. Wu, H. Zhou, X. Shen, Dynamic cell association for non-orthogonal multiple-access v2s networks. IEEE J. Sel. Areas Commun. **35**(10), 2342–2356 (2017)
9. X. Lian, Y. Huang, Y. Li, J. Liu, Asynchronous parallel stochastic gradient for nonconvex optimization, in *NIPS* (2015)
10. X. Lian, W. Zhang, C. Zhang, J. Liu, Asynchronous decentralized parallel stochastic gradient descent, in *International Conference on Machine Learning* (PMLR, 2018), pp. 3043–3052
11. S. Wang, T. Tuor, T. Salonidis, K.K. Leung, C. Makaya, T. He, K. Chan, Adaptive federated learning in resource constrained edge computing systems. IEEE J. Sel. Areas Commun. **37**(6), 1205–1221 (2019)
12. W. Zhang, S. Gupta, X. Lian, J. Liu, Staleness-aware Async-SGD FOR distributed deep learning, in *IJCAI* (2016)
13. A. Reisizadeh, A. Mokhtari, H. Hassani, A. Jadbabaie, R. Pedarsani, Fedpaq: a communication-efficient federated learning method with periodic averaging and quantization, in *International Conference on Artificial Intelligence and Statistics* (PMLR, 2020), pp. 2021–2031
14. M.M. Amiri, D. Gündüz, Machine learning at the wireless edge: distributed stochastic gradient descent over-the-air. IEEE Trans. Signal Process. **68**, 2155–2169 (2020)
15. A. Shokrollahi, Theory and applications of raptor codes, in *Mathknow* (Springer, Berlin, 2009), pp. 59–89
16. M. Luby, M. Watson, T. Gasiba, T. Stockhammer, W. Xu, Raptor codes for reliable download delivery in wireless broadcast systems, in *CCNC 2006. 2006 3rd IEEE Consumer Communications and Networking Conference, 2006.*, vol. 1 (IEEE, Piscataway, 2006), pp. 192–197
17. P. Auer, N. Cesa-Bianchi, P. Fischer, Finite-time analysis of the multiarmed bandit problem. Machine Learn. **47**(2), 235–256 (2002)
18. S.U. Stich, Local SGD converges fast and communicates little (2018). arXiv preprint arXiv:1805.09767
19. F.N. Iandola, S. Han, M.W. Moskewicz, K. Ashraf, W.J. Dally, K. Keutzer, Squeezenet: Alexnet-level accuracy with 50x fewer parameters and <0.5 MB model size (2016). arXiv preprint arXiv:1602.07360
20. V. Patel, Kalman-based stochastic gradient method with stop condition and insensitivity to conditioning. SIAM J. Optim. **26**(4), 2620–2648 (2016)
21. M. Jaggi, V. Smith, M. Takáč, J. Terhorst, S. Krishnan, T. Hofmann, M.I. Jordan, Communication-efficient distributed dual coordinate ascent (2014). arXiv preprint arXiv:1409.1458
22. L. Bottou, F.E. Curtis, J. Nocedal, Optimization methods for large-scale machine learning. SIAM Review **60**(2), 223–311 (2018)

Chapter 6
Conclusions and Future Workers

In this chapter, we conclude the monograph with the overall summary, and provide some promising directions regarding the Internet access of vehicles and novel architectures for further development of new generation IoV Internet access mechanisms.

6.1 Conclusions

We have investigated the Internet access for vehicles, mainly focus on the V2R access method over unlicensed spectrum resources. To effectively setup Internet connection, there are three main steps that a vehicle must accomplish before real transmitting. We have analyzed the access delay and throughput performance using the roadside access points. After that, we have discussed the data traffic offloading problem given the coexistence of V2V and V2R data pipes. Both analytical model and experimental research are proposed to utilize the interworking of V2X communication. Besides, we have studied the link management of the V2R communication, and employed machine learning based algorithms to improve the rate control to adapt to the highly dynamic channel condition. In addition, we have also investigated to apply the distributed learning among vehicle users, and propose two asynchronous training methods. The main content of the book is presented and summarized as follows.

(1) Internet access modeling to evaluate the delay and throughput performance. Providing Internet access to vehicle users is of great importance. We consider the actual requirements for vehicles to setup effective V2R access to connect to Internet services. The access procedure is modeled as a Markov chain by defining the system status that rely on the transmission of a set of management frames. To better understand the Internet access procedure, we have provided analytical framework to find out the relationship between the access perfor-

W. Xu et al., *Internet Access in Vehicular Networks*,
https://doi.org/10.1007/978-3-030-88991-3_6

mance and various factors including channel status, access protocol, velocity, etc. Such results is verified by the conducted experiment, which are very useful for future vehicular access protocol design, especially with high mobility scenarios.

(2) V2X interworking for traffic offloading for Internet access. We have investigated the possibility to utilize the economical RSU to offload the data traffic from expensive cellular networks. An M/G/1/K queueing framework is setup to analyze the tradeoff between the offloading efficiency and transmission delay. Moreover, we have further analyzed the V2V assistance for improving the vehicular offloading performance based on the interworking of V2R and V2V communication.
(3) Intelligent link management for vehicular Internet access. To adapt to the high mobility of vehicles, it is essential to track the variation of the V2R channel and adjust the link configuration accordingly. We have employed machine learning based methods, i.e., deep learning based classifier, reinforcement learning, to enable the optimal MCS selection to improve the throughput of the wireless link between an access station and a mobile user, including vehicles, UAVs, maritime vehicles, etc.
(4) Distributed learning over Internet access of vehicles. We have considered two different access methods, i.e., a macro cell covering multiple vehicles within a relatively large area, and a roadside cell that can only opportunistically communicate with the drive-by vehicles. The proposed two learning paradigms can show that efficient distributed training for ML models can be achieved with convergence guarantee.

6.2 Future Directions

In this monograph, we have presented the Internet access for vehicle users from a practical perspective based on analytical modeling and empirical evaluations. There are still several interesting research directions including new Internet access protocol design and development as well as the interplay between the ML framework and new IoV networking architectures. Although significant efforts have been put on the synergy of ML and wireless networks, however, some issues related to the specific requirements and unique characteristics of IoV remain open. For example, how the mobility of vehicles affects the distributed learning patterns, and what the potential data path is brought by the location changing of users, etc. The following topics along the way of this monograph are worth to be investigated.

(1) Interplay between IoV and machine learning. Status quo literature often utilize ML as a tool to optimize the vehicular network performance, and has seldom considered the mutual effect between networking and machine learning applications. To better understand how the two can help to improve each other, it is meaningful to conduct systematic investigations on the interplay between

them, i.e., to seek potential schemes that can optimize the communication performance and boost the machine learning procedure simultaneously. For example, the IoV communication trace can be mined to support big data acquiring, transmission and processing which can benefit both the model training and inference, and thus can advance a variety of intelligent automotive applications.

(2) Distributed training architecture for large scale vehicles. Currently, the training tasks are deployed over vehicles, edge stations or cloud centers. To utilize the valuable features in vehicular network such as the mobility of the connected vehicles which have strong computing capability and significant storage space, virtual information pipes can be set up between road users since vehicles can carry extremely large files from one place to another within much less time than using status quo network infrastructure, and can be possible used for neural model exchanging, big data transmission, etc. Such virtual information pipe can help to accelerate the distributed training process by considering new communication and computing architectures, such as asynchronous aggregation, fountain code based broadcasting, etc.

Printed in the United States
by Baker & Taylor Publisher Services